# Practical Engineering Design

# Practical Engineering Design

Edited by
## Maja Bystrom
## Bruce Eisenstein

Taylor & Francis
Taylor & Francis Group

Boca Raton London New York Singapore

A CRC title, part of the Taylor & Francis imprint, a member of the
Taylor & Francis Group, the academic division of T&F Informa plc.

Published in 2005 by
CRC Press
Taylor & Francis Group
6000 Broken Sound Parkway NW, Suite 300
Boca Raton, FL 33487-2742

© 2005 by Taylor & Francis Group, LLC
CRC Press is an imprint of Taylor & Francis Group

No claim to original U.S. Government works
Printed in the United States of America on acid-free paper
10 9 8 7 6 5 4 3

International Standard Book Number-10: 0-8247-2321-X
International Standard Book Number-13: 978-0-8247-2321-7

**Library of Congress Cataloging-in-Publication Data**

Catalog record is available from the Library of Congress

Taylor & Francis Group
is the Academic Division of T&F Informa plc.

Visit the Taylor & Francis Web site at
http://www.taylorandfrancis.com

**and the CRC Press Web site at
http://www.crcpress.com**

# PREFACE

Design gives engineering students an opportunity to practice their profession by designing a real-world solution to an engineering problem in a way that can best be likened to an apprenticeship. In the early days of engineering education, virtually the entire curriculum consisted of students working with faculty and practitioners from industry in an apprenticeship setting to solve problems that were identified by local industry as critical. Thus, many engineering schools began as mining, agriculture, or technical colleges to serve local needs.

Following World War II and the enormous number of technological innovations spawned during the war years, engineering schools shifted to an *engineering science model* of engineering education in which all courses became science based instead of experience based. Although the pendulum is swinging back in the direction of an experience-based engineering education, today's students are left with practically no exposure to the real problems facing industry and society if they do not have a formal program in engineering design.

The main benefit of the design project you are about to undertake is the opportunity to recapture the practical experience that is now regrettably missing from the rest of the engineering curriculum. Students can have the satisfaction of carrying a project through from the early design stage to a working prototype. In the process, you will learn about working in teams, scheduling, budgeting (both time and dollars), fabrication, documentation, and presentation. You will have the pride that comes from a job well done and, as an added bonus, an entry on your resume that will help in getting a good job. For those who envision their future in an entrepreneurial track, the experiences and knowledge that you will gain from a well-done design project can accurately mirror the process of running your own company.

At the outset, a design project is intimidating: you are required to either take a project description from an industry or faculty advisor or come up with your own project idea and find an advisor. You then must complete the project design and documentation and turn in a completed prototype in a relatively short amount of time. It is no wonder, when you have very little practical training, that this may seem to be a difficult, if not impossible, task. However, a successful design project is certainly achievable if you break the project into subtasks, turn to faculty or industry advisors for assistance, and keep yourself and your team organized.

Most universities have some form of engineering design in order to allow students practice in completing an entire extensive project. A formal course gives you the room to explore and experience the design process at relatively little risk (other than a low grade) before you are plunged into your first real job and are entirely responsible for the development of a product or design.

Even if you have had solid work experience through co-op jobs or other means, a formal course in design gives you the opportunity to hone the skills necessary for success in any work position. In both course-based design and real jobs you will typically work in a group, and therefore you will have to be able to successfully allocate duties and resolve intra-group conflicts. A design project gives you the chance to experience working in a team whose members have different working styles and personalities, with an advisor available to help if there are any difficulties. The design process will also help you to hone your engineering skills, the problem-solving skills that have been subtly inserted into all of your classes since you were a freshman. In this way it brings together everything you have learned — not the obscure mathematical transforms or chemical processes required in undergraduate classes but rather the thought processes required to arrive at conclusions from a starting set of ideas or hypotheses by examining the trade-offs and alternative paths. Finally, it is almost impossible to be a good engineer today without possessing good communication skills. Unfortunately, oral and written communications are among the topics that are often overlooked or underemphasized in engineering curricula. By producing written (and often oral) reports of your project, you can obtain immediate feedback and learn how to fine-tune your presentation skills. This will benefit you immensely when you take a job in industry and are required to write project proposals or present ideas to your managers. While it is often the case that the number of requirements for a design project are great, and the allotted time very short, it is an almost unanimous opinion of graduating seniors that the design experience was valuable and, in most cases, enjoyable.

This handbook is an outgrowth of a set of senior design guidelines used in the Electrical and Computer Engineering Department, and as an extension of the *Senior Design Handbook* written in 1996 by James E. Mitchell for the Civil and Architectural Engineering Department at Drexel University. This handbook is aimed at students in design classes as well as novice engineers who are taking on their first project as a co-op assignment or on their first permanent job. It was written to answer some of the most pressing and most often repeated questions that we have heard from seniors or novice engineers as we have either taught various courses on design or served as project advisors. Unfortunately, there is no formal checklist by which you can complete a design process; however, this handbook provides a set of guidelines and includes examples of different aspects of the process. This is by no means a complete manual for the design process, and thus at the end of every chapter we have included what we feel are some of the best references on the topics covered. We strongly urge you to turn to these references for further information and simply use this book as a set of guidelines for your project.

The topics are arranged in the order in which you are likely to encounter them during your project. In the first chapter an overview of the design process is given. If you are at a loss as to where to start your project, you should read this chapter in detail. If, however, you already have a solid idea in mind, have formed a team, and have begun the design, this chapter will refresh your memory of important design aspects.

The second chapter briefly discusses how to consider the impact of your project on society. Often, the ethics and social impacts of projects are subtle and require considerable reflection to discern. The chapter provides a series of questions that you and your team members can ask yourselves to help determine the benefits and risks of your project and can serve as a guideline as you move from the world of academia into industry after graduation.

The third, fourth, and fifth chapters cover project scheduling, management, and budgeting. While these chapters contain useful tips on these topics, not all of the material presented will be required for all design projects. For instance, in Chapter 5, three types of project budgets are discussed. Your design course may only require that you produce a budget for your prototype. However, it may be interesting for you to scan the remainder of the chapter to see how budgets might be produced for a project in industry.

The sixth and seventh chapters contain guidelines for written and oral presentation of your work and discuss how to make your project appealing to a variety of audiences. Obviously there is no clear-cut formula for doing so; much effort has to be put into writing and speaking, and many revisions performed to obtain fine-tuned results. These chapters provide important, but by no means exhaustive, tips for arranging the written documents and the oral presentation.

The eighth and ninth chapters are included as extra material for those who have unique projects with intellectual property that should be protected or that can be expanded into a business. Again, these chapters are not complete discussions of the topics, but are designed to answer basic questions such as what is intellectual property and what steps would be needed if you wanted to use your ideas to start a business. If you think your ideas should be protected, we encourage you to contact your organization's intellectual property office (most companies, colleges, and universities have such an office, typically as part of their research offices, or at least have a full- or part-time lawyer who can advise on these issues). If you are a student and think you may want to start a business based on your project (we know many students who have done so), you should contact the entrepreneurial or incubator center at your university. If your university does not have facilities such as these, you should consider finding a faculty member who has experience in starting a company and ask for ideas or assistance. It is also a good idea to take a business or management course or two so that you will know what to expect as you start to deal with the complex issues of finding financing and managing employees.

We have included in the Appendix a set of excerpts from senior design reports and presentations that you can refer to in order to obtain ideas about style and content of different documents. These documents were selected to represent different types of projects and are what we consider to be both excellent selections of project topic and well-written documents. These reports will be referred to throughout the document and both their qualities and faults will be discussed. The faults and omissions of the reports were not necessarily due to laxness on the part of the authors; they were in part due to the changing requirements as the Senior Design course evolved at Drexel University. These documents emphasize the fact that no matter how well designed and presented, a project and its documentation can always be improved. We thank the authors of these reports: David Brouda and Kevin Lenhart for *The Talking Book* and Keith Christman, Adam O'Donnell, Chayil Timmerman, and Suma Varghese for *Coreware IPv4 to IPv6 Bridge*. Without their generosity in sharing their work, this book would have been missing a vital section.

You will find that there are some points that are emphasized over and over again, or repeated through every chapter (such as getting an early start, keeping backups and good notes, asking people for help, taking care to consider all aspects of the design, documentation, and presentation). This is not because there is a lack of topics to cover, but rather because all of the authors have learned through experience how important these points are.

Finally, as mentioned above, this handbook is not an exhaustive list of design procedures and requirements and does not contain solutions to all of the problems and challenges you will encounter during your design project. You should work with your manager, advisors, or mentors such as other faculty in your department or senior staff in your company to ensure that you are meeting the requirements of the course and have not omitted significant steps in the design process. You should also consult your peers and ask for candid opinions of your project ideas and presentation, since (just as in a mechanical or electronic system) feedback is often the most valuable tool in improving the output. Often it only takes another perspective to obtain insight into the solution of a difficult problem, or your peers may know of someone who can assist you. You should also consult the books and Web sites in the suggestions for further reading, since they contain more complete discussions of the topics.

We wish you success during your design project. With time, care, and attention, everyone can produce an excellent design or prototype while gaining experience in the many aspects of the design process.

**Bruce Eisenstein**
**Maja Bystrom**

# CONTRIBUTORS

**Valarie Meliotes Arms**
Department of English and
  Philosophy
Drexel University
Philadelphia, Pennsylvania


**Kimberly S. Chotkowski**
InterDigital Communications
  Corporation
King of Prussia, Pennsylvania

**Moshe Kam**
Electrical and Computer Engineering
  Department
Drexel University
Philadelphia, Pennsylvania


**Robert J. Loring**
School of Biomedical Engineering,
  Science and Health Systems
Drexel University
Philadelphia, Pennsylvania

**James E. Mitchell**
Department of Civil, Architectural,
  and Environmental Engineering
Drexel University
Philadelphia, Pennsylvania


**Stewart D. Personick**
Electrical and Computer Engineering
  Department
Drexel University
Philadelphia, Pennsylvania

# CONTENTS

## 5   Are We in Business Yet? ...................................................................79

*Stewart D. Personick*

## 6   Documenting Your Design Project......................................................97

*Maja Bystrom*

## 7    Presenting Your Design Project ...............................................135

*Valarie Meliotes Arms*

## 8    Intellectual Property ...............................................................149

*Kimberly S. Chotkowski*

# APPENDIX

★ *Key Points*
*Chapter 1*

- Design is what engineering is all about. The courses, the laboratories, the analysis tools, and in fact your entire undergraduate engineering education, are preparation for you to learn how to undertake a design project. For most engineering students, the first exposure to design is the capstone sequence senior design or the first internship or co-op position with a company.

- A senior design course sequence gives you the opportunity to experience a real-world design process in a laboratory setting. As a consequence, you can take on an ambitious project with little actual risk. Similarly, your first design project for industry will likely be as a member of an experienced team and will be well supervised by your manager or team leader. Often, there are formal, written design procedures and practices that will provide the framework for your project.

- There are eight primary steps in a design process:
  - Forming your team and recruiting an advisor
  - Identifying an opportunity or a problem that needs an engineering solution
  - Selecting and evaluating a design project within the scope of the opportunity or problem
  - Generating and evaluating design alternatives (including economic consequences)
  - Modeling and simulating one or more designs or components
  - Selecting and implementing one design
  - Testing and verifying your design
  - Documenting and presenting your project

- By taking care to complete each step you will minimize problems along the way and have a successful project.

# Chapter 1

# THE DESIGN PROCESS

*Bruce Eisenstein, Stewart D. Personick,
James E. Mitchell, and Maja Bystrom*

## 1.1 WHAT IS A DESIGN PROJECT?

Engineering design projects are as varied as the opportunities and problems within our society. The choice of projects will be influenced greatly by the interests of the person or team working on the project, the requirements of the company, and even by the engineering disciplines represented on the team. Interdisciplinary projects add another dimension and can be the most interesting and rewarding of projects, since they give all team members an opportunity to learn about different fields. In the Appendix we have included examples of two vastly different electrical engineering senior projects. Each had a different set of requirements, deliverables, and, as a consequence, necessitated different team skills. The following are the titles of some recent projects at various universities:

- Multimedia Filtering
- Desktop Integrated Microcontroller Environment (DIME)
- Bird Sound Recognition Algorithm
- The Robotic Snake Comes of Age
- DSP Laboratory
- Solar Converter
- Battery Sensing and Monitoring

From this list you can see that even in one engineering discipline there is no typical project, and you can see that each of these projects would involve different techniques and result in different deliverables.

In general, a design project can result in a new service, product, process, or design. The common thread that is shared among all disciplines is that the project has to be unique and it has to involve elements of design. That is, you must design or create something new. This does not necessarily mean that you need to design a method, process, or consumer product that has never been designed before. You may choose to redesign a current process or product, but, when you are finished, there must be something valuable added by your work that was not there before.

All projects, whether performed as course work or for industry, will be graded in some sense. For instance, the course project will result in a letter grade for the class and, perhaps, later recommendations from faculty advisors, while the industrial project outcome might have a significant impact on an annual review and merit raise. Your project will be graded or rated on the basis of how well you and your group adhere to the schedule and deliverables that are in your proposal. Since you will be working under a time limit, it is important that your proposal clearly delineates what you expect to accomplish and by when. A design project can be compromised by being too ambitious, causing time or budget overruns, or at worst an incomplete delivery. It is much better to scale down the proposal so that the deliverables are attainable, and at the time of the final presentation a finished and tested product is there for the evaluator to see. A very well-done, finished, and tested project is better than a more ambitious project that is incomplete.

A good rule to follow is that the best projects are those that meet the customer's needs and expectations. The customer may be a manager, an advisor, the department faculty or staff members, a senior design committee, or a company client. These customers may be providing funds for the work or may be relying on your project in order for another project to continue. Therefore, they will be mainly interested in having their specifications met within the given timeframe. However, in addition to the deliverables, the customer will be interested in ongoing maintenance costs, ease of manufacturing, and prospects of satisfying environmental considerations.

## 1.2   THE STEPS TO A SUCCESSFUL PROJECT

There are eight (not necessarily sequential) steps required for completion of a design project:

- Choosing your team and advisor
- Identifying an opportunity or a problem that needs an engineering solution
- Selecting and evaluating a project within the scope of the opportunity or problem
- Generating and evaluating alternative design solutions and deciding on a "figure of merit" against which to compare the different approaches
- Modeling and simulating component interactions or system performance
- Selecting and implementing one design approach
- Testing and verifying your design
- Documenting and presenting your project

These steps are not sequential and can be implemented as in Figure 1.1. There are several observations that you should make from this figure. First, design is iterative. You must evaluate and test design alternatives, and often you must go back and redesign a system. Second, at each stage of the design, you must document your work. Finally, some of the steps (those indicated by dashed lines) may be optional. You may have your team, project, and detailed specifications assigned. If so, you are ready to start the project and can immediately jump ahead to the section on generating design alternatives.

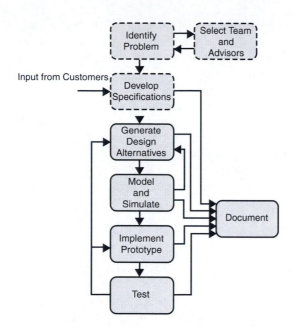

**FIGURE 1.1**
The design process.

If you are responsible for all or part of team, if you are the team advisor, or if you are still engaged in the process of project selection, read on. In this chapter we discuss each of these steps, and in subsequent chapters we provide more information on the design process.

### 1.2.1 Choosing Your Team and Advisor

In industry, almost all design projects are performed in teams, usually assigned and not selected by the participants. An important part of the education you will get in a formal design course is the ability to work as a team. While a team may draw on many individuals throughout the course of a design project, the key people are the teammates and their advisor. Since this handbook is aimed at novice engineers, it is likely that you will need or want an advisor for your first design project. If the project is for a required course, your advisor might automatically be assigned or you might have a list of faculty members to choose from. If this project is one of your first industrial design projects, your manager can serve as an advisor, and you might seek additional advisors and mentors, if they are not assigned by your manager. Wisely selecting and carefully considering the needs of the project and the skills of each of these individuals can make an enormous difference in both the ease and the pleasure with which you move through the entire process. The following suggestions should assist in making those selections effectively.

#### *The Team*

The first inclination is to pick friends from your class or your department as your team members. Often this can work well, especially if they are study partners or teammates whom you have worked with before. Selecting them solely because they are friends will help the project as much as selecting them for their partying ability or because they work in the neighboring office.

Picking the best team depends in large part on weighing three important criteria. First should be the nature of the design you're undertaking. You want individuals with the skills and previous experience necessary to complete the work or those who are interested in and have the ability to obtain the necessary skills. Second are the personalities of each of the members. You probably want both a somewhat impulsive "get-it-done" person and a more restrained "let's-think-about-it" person to balance each other. If this were a space flight you would have undergone extensive psychological testing in order to determine compatibility. As it is, you need to at least think carefully about how you will get along together for many long months. Team roles will be discussed in Chapter 4; you should consider ahead of time whether the teammates you are choosing are able to span roles from team leader to organizer to recorder, since you do not want a team in which all members wish to be leaders. Third are the schedules and other commitments of all the team members. To be successful you will need to see each other regularly as a complete group. Be sure you are able to do so.

### The Advisor

Picking the right advisor is as important as picking the right team, and worth considering equally carefully. For course-work projects, you probably know the potential advisors because you have had them in class or you know them by reputation. Personal knowledge is a good indicator of how they will work with you as a team advisor, but bear in mind that working on a project is very different from taking a class. Therefore it is incumbent on the group to discuss with potential advisors what the advisor expects of a team in terms of level of input and communication frequency before officially signing on the person. Some pitfalls that would work against a particular advisor are: schedules that are incompatible with your group's, expectations of a level of work that is either greater or less than you envision, requirements for progress reports, and evaluation methods or other specifics that are out of line with your team's wishes. It is important to understand these issues as well as possible in advance and discuss them among the team members in order to make an informed decision.

One particularly useful question to ask is how potential advisors envision their role working with you. Some advisors may see themselves as project leaders, with you taking the roles of helpers carrying out their vision. Others may wish to act solely as consultants, waiting for you to ask questions. Many advisors can tell you in advance what they prefer, thus reducing the uncertainty of what the working relationship will be like. Yet another question worth discussing in advance is how often they will be able to meet with you. The average for a design course is once a week, but some advisors are so busy that this frequency is not possible. In industry, some potential advisors may be more flexible or busy than others. Some may have open door policies — if the door is open, then walk in and ask — whereas some may require that you make an appointment for consultation. Again, knowing in advance what to expect will help you during the project.

Industrial advisors come in many forms. The first kind is the team members' manager, or for cross-departmental teams, the managers of the associated groups. The second potential kind of advisor is a senior staff member who has worked on similar projects in the past. Both types of advisors would know the company's expectations for timelines, budgets, and report generation frequency and format. Also, both types of advisors, but particularly the latter, can be especially useful in networking, that is, in finding other people in the company who can assist you with your project. To find a senior staff member to assist as an advisor, you should ask your manager or other senior staff you trust for opinions on potential advisors and for introductions to these people.

Often faculty advisors for course-work projects come with specific topics or projects in mind. Your department may list faculty or outside projects and the associated advisors, specifications, and funds. In this case, if you choose to work on a project and the faculty member or outside sponsor agrees to enlist your aid, then your advisor is fixed, and you must learn to work with him or her. In some cases, the project and advisor will be assigned to you and you have little, if any, choice in the matter. It is still a good idea to meet with advisors before the project starts and understand what they expect of you and what you should expect from them. If you have any sense that the advisor might not be a good fit for you, it would be a good idea to either choose a different topic and advisor or request an advisor change for an assigned topic. In all of these cases you should also consider finding a co-advisor.

### Co-Advisors

Co-advisors, either from the university faculty or from industry, are typically not required, but may be very helpful on some projects. They usually act as consultants for particular technical areas whenever a need arises. For instance, you may need the aid of someone who is an expert in software engineering if your advisor is only an electronics expert or you may need the help of a technical writing expert if you know your team is particularly weak in that area.

It is usually not necessary to sign co-advisors up officially at the beginning of a project unless you know that extensive advice will be necessary. What you may do is ask them at the beginning of the project whether they would be available to help; if they will be unavailable for certain times, you should look for others who can provide input into needed areas.

### 1.2.2  Identifying an Opportunity or a Problem

If you already have a topic in mind or have been assigned a project by your manager or course coordinator, then you can proceed to Generating and Evaluating Design Alternatives. However, thinking about the problem you are solving and evaluating topics might help you define your product specifications and develop a solution. If you must find your own project, then you first must identify an opportunity or problem. Then, after discussion with your team or advisors, you will be able to turn the problem into a project and ensure that you will be successful. The difference between a problem and a project topic is that you must evaluate the project topic to see if it is doable, and whether it is really a good idea. You must be prepared to evaluate the ideas fairly and to reject your topic if the evaluation results determine that it is not feasible.

If you are given a free hand in selecting your project topic, then while you are finding team members and an advisor you should start to find a project topic. Project topics come from many places. Often faculty within and outside the department will list topics of interest to them. For senior design classes your department senior design coordinator may also solicit topics from local businesses, organizations, or individuals. Excellent ideas often come from contacts students make during internships or co-op assignments. Some departments will require that students develop their own project ideas and may provide a mechanism for students in the senior design course to share ideas and form teams of people with complementary skills and ideas. Typically, the most interesting and rewarding projects are those that solve a real-world problem — those that will have lasting effects on your customer. If you are choosing a project from a list or on the advice of a co-op manager, make sure that the topic is of interest to you and that there is an audience for your solution, that is, people who will make use of it.

Opportunities and problems that are good candidates for design projects are all around us: a slow elevator that stops too long on each floor, parking meters for which you never seem to have the right change, inefficient use of space in a building, need for a better medical diagnostic test, desire for improved energy efficiency and reduced energy consumption, ways of making the environment better or health better, in fact, any mitigation of pain, suffering, or environmental damage, and general improvements of life on Earth.

Since you are surrounded by opportunities and problems, you would think it would be easy to come up with a good project topic. However, design teams seem to find it difficult. The reason, we believe, is that novice designers often fall into the trap of deciding on a solution before they have selected the problem. They are then in the position of a mechanic who selects a tool and then looks for a place to use it. For instance, students will come to us with a request to do a senior design project that uses Fourier analysis or makes use of a new buzz word technology like fractals or a new composite material. These, however, are tools, and not problems. Projects that do not start with an opportunity or a problem usually do not turn out well.

There is an interesting and important element of timing in selecting a problem or opportunity to address. There are problems that would be great to solve but for which there are no known solutions. There are technologies that would be nice to apply but for which there are no useful problems to apply them to. Real opportunities to create useful engineering solutions occur when one can identify, *simultaneously*, a problem that is worth solving and a potential engineering solution. Applying integrated electronics to create a practical personal computer was a great idea in 1975. Making a coffee cup insulator out of corrugated paper was a great idea in 1999. Such opportunities are usually fleeting, because others will also identify them.

### 1.2.3   Selecting and Evaluating Your Project Topic

#### *Selecting a Topic*

Starting with a list of problems you might wish to address, spend time brainstorming, by yourself, with coworkers, with friends, with family members, and with your group, if it is formed. Your goal in this stage of the process is to develop a set of possible approaches to solving each problem. Later you will narrow the set down to feasible and useful approaches.

Ask people how they might approach solving your selected problems. Keep a list of all ideas; do not discard any, even if they seem uninteresting or outrageous. Revisit the ideas on the list occasionally. There is usually something salvageable from even the more outrageous ideas, and often initially uninteresting topics turn out to have interesting features that, with some negotiation, can be turned into an outstanding project. If you find yourself lacking ideas, try restating the problem in a different form. As will be seen below, looking at the problem in a different manner often leads to ideas.

---

▼

---

#### STATING AND RESTATING A PROBLEM

One example of a problem is overly-hot take-out coffee served in paper cups.

You might start your brainstorming sessions by phrasing the problem in exactly this way: "The coffee is too hot." This might lead to several possible solutions (some of which are simpler or more feasible than others and some of which may not be suitable for a short-term design project):

- Ask the store to reduce the coffee machine's temperature.
- Request that ice be served with the coffee.
- Design a mug that chills the coffee to the desired temperature.
- Build an "instant chiller" — the opposite of a microwave oven.

You might also rephrase the question or make it more specific, that is, "The coffee cup is too hot to hold." This might lead to a different set of solutions:

- Design an insulating cup.
- Design a part of the cup that is insulated.
- Build a cup holder into a car or a piece of clothing or a briefcase.

Naturally there are other solutions to the problem, and it could be restated in different ways that might lead you to them.

━━━━━━━━━━━━━━━━━━━━━━━━━▲━━━━━━━━━━━━━━━━━━━━━━━━━

After you narrow the field down to one or (hopefully) more solutions, but before you begin a more formal evaluation of each possible project topic, ask others for their opinions on the project and take these seriously. As you discuss these projects with peers, teammates, and advisors, you must also keep evaluating the ethical issues involved with potential projects. Chapter 2 gives a list of questions you can ask yourself while you are considering topics and evaluating alternative solutions, but it is also useful to discuss possible ethical ramifications with people who may have different perspectives on the project than you.

## Evaluating Your Topic

After narrowing your list down to several choices, it is necessary to start a formal evaluation process. Much of this process will be the background research that will form the basis of your motivation, budget, and time line, so it is not an idle process, and the sooner it is begun, the sooner you will be able to continue with the "meat" of your design project. If the topic does not pass this evaluation step, you must return to the earlier steps of selecting a new topic and possibly even selecting a new problem.

Engineering is a creative process that produces goods and services that people value and trust. However, regardless of the societal benefit, or the contribution that an engineering endeavor may make to the knowledge base of science and technology, one must remember: engineering projects are funded by investors. The investors may be venture capitalists, family members, a government organization, or the shareholders of an existing company. Those investors expect a return on their investments. For commercial projects, the return on investment is profit.

Luck plays a significant role in the success of new companies, new products, and new services, but there are things you can do to significantly improve your chances of success. More importantly, there are things you can do (or fail to do) that will almost guarantee your lack of success.

To have a successful commercial project you must ask yourself all of the following questions:

- What is the product or service that we will create or produce?
- What is the unmet market need that is addressed by our product or service?

- What products and services already exist that serve, at least in part, this market?
- What is new and different about our idea?
- Why do we think we can make it work?
- How will we reach our customers?
- What are the known risks?
- How hard will it be for a competitor to take our market share away?
- Is there a threshold or barrier to entry for competitors?
- What do we have to do to compete successfully in the marketplace with this product or service?
- How much do we have to invest, and how long will it take before we become profitable?

These questions are a lot easier to ask than they are to answer, but failure to think about these questions, and their possible answers, significantly decreases your chances of success. For this part of the design process, you are effectively developing a business plan, that is, building a case for a product or design that would be needed or desired by a customer. We will focus on the first seven questions, because these must be answered to some extent before you settle on your final topic.

### What Is the Product or Service that You Will Create or Produce?

One of the most common mistakes made by people who are thinking about bringing a new product or a new service to the marketplace is that they do not understand what their product or service is or how it fits into the "value chain" of a larger product or service.

------------------------------▼------------------------------

### THE VALUE CHAIN: AN EXAMPLE

Alice and Bob have an idea for a new service: satellite delivery of radio programming to people in automobiles. The concept is to provide people with more reliable reception, particularly in rural areas. One of the things that Alice and Bob need to think about is the value chain associated with this new service concept. A partial listing of the links of the value chain are:

- Production of recorded music by artists and studios
- Production of news and talk show programming
- Acquiring the rights to receive, repackage, and redistribute radio programs (music, news, talk shows) that are produced by others
- Delivery of music, news, and talk show programming to satellite ground stations
- Design of satellites and satellite ground stations
- Acquisition of property for installation of satellite ground stations
- Acquisition of radio spectrum for use by satellite radio services
- Installing ground stations
- Launching satellites
- Operating satellite systems
- Leasing and reselling satellite and ground station capacity
- Designing radios capable of satellite radio reception

- Producing radios capable of satellite radio reception
- Selling radios capable of satellite radio reception
- Installing radios capable of satellite radio reception (in automobiles, boats, etc.)
- Selling advertising time to businesses that wish to advertise on satellite radio
- Charging an adequate subscription fee to cover costs and ensure a profit

Even though the overall satellite radio service concept may be a good idea in terms of the desire of people who have automobiles and boats to purchase such radios and the desire of advertisers to purchase advertising time on this service, there is still the very important question of what role Alice and Bob want their company to play in the value chain.

It is not likely that they want to get into the business of producing recorded music. They probably are not thinking about getting into the business of designing and launching satellites. They probably are not going to start a company to manufacture radios capable of receiving satellite radio (because existing, giant consumer products companies already do things like that very effectively). Maybe they want to get into the business of leasing capacity from companies that own and operate satellites and ground stations and reselling that capacity to radio networks that wish to deliver their programs using satellite radio. Maybe they want to get into the business of selling, installing, and servicing satellite radios that are manufactured by other companies.

▲

▼

## The Value Chain: A Second Example

Charlie and Donna have an idea for a new PDA (personal digital assistant)–based application that allows people to access stored data, a feature that is not easy to achieve with today's products. Even if they are convinced that there will be a demand for this capability, they still have to decide what role they will play in bringing this concept to life. It is unlikely that Charlie and Donna want to get into the business of manufacturing special purpose PDAs (because the PDA market is already served by some very big, established companies and because people do not want to carry around more than one PDA). Charlie and Donna might want to develop and sell "client" application software that runs on general purpose PDAs; or perhaps they might want to sell custom-designed software and associated installation and support services to companies and organizations (such as your University) that wish to make portions of their stored data available to end users with PDAs.

▲

## What Is the Unmet Market Need that Is Addressed by Your Product or Service?

To be successful, the product or service you sell must fill an unmet market need. That is, there has to be some motivation for your target customers to buy your products or services. The unmet market need could be for a less expensive or an easier-to-use alternative to an existing product or service. The unmet market need could be for:

- Higher performance
- Lower cost
- Greater ease of use
- Less required maintenance
- Higher durability
- Improved customer service
- Increased convenience

The unmet market need could also be for an entirely new product or service, for which there is no reasonable alternative currently available.

---

### ▼

### UNMET MARKET NEEDS

A new audio system design might produce improved sound over existing systems, for example, higher performance.

A wireless local-area-networking product might be more convenient to use than a wire-based solution.

A semiconductor storage device might be more resistant to vibration than a magnetic disk–based device (or require less maintenance or have higher durability).

A computer that accepts spoken commands might be easier to use than a computer with a keyboard and mouse–based user input interface.

### ▲

---

It is difficult to think of a new product or service for which there is no reasonable current alternative serving the associated market need. The invention of the telephone was revolutionary, but it could be (and was) viewed as a new alternative to the telegraph and messenger boys.

The case for a new product or service must be clear and convincing with respect to the unmet market need (or set of needs) it is going to address. It might turn out that there are valuable uses for your product or service that you never dreamed of. It might turn out that these unanticipated uses produce far more sales revenue than the unmet market needs you thought your product or service would address. Nevertheless, if you do not have a clear vision for the unmet market need that your product or service is going to address, your chances of success are greatly diminished. It is unlikely that any investors will invest money in a new product or service that does not address a clearly identified target unmet market need.

### *What Products and Services Already Exist that Serve, at Least in Part, this Market?*

Having identified (at least tentatively) the unmet market need you plan to address with your new product or service, you must carefully consider the existing products and services that can, at least in part, serve that same market need.

▼

## EXISTING SERVICES AND MARKET NEEDS

The Iridium low-earth-orbiting satellite system was conceived and implemented to allow business people traveling in rural and underdeveloped regions of the world to make and receive telephone calls. Unfortunately, in the same time frame that Iridium was being created, existing terrestrial cellular systems were being widely deployed in essentially all places in the world where one might want to do business. These terrestrial cellular services are a less costly and often more reliable solution for serving that market need. Although widely deployed terrestrial cellular services did not exist at the time the Iridium system was being conceived, one might speculate (with hindsight) that the widespread deployment of terrestrial cellular systems should have been anticipated by a consortium of companies preparing to spend billions of dollars on a new telecommunications service and the associated satellite and ground station infrastructure. The much higher cost of an Iridium phone call priced them out of the market for all but a very limited number of users. The original Iridium consortium went bankrupt, and the Iridium satellite system was, subsequently, purchased by another group of investors. There are a number of suggestions being entertained about how to use the existing satellites and ground facilities.

▲

The existence of alternative products or services that serve, in part, the unmet market need you plan to address does not mean that your product cannot succeed. However, those who fail to consider such alternative solutions greatly reduce their probability of business success.

Every business case should include a thorough analysis of alternative (competing) solutions for serving the market need that is targeted by the proposed new product or service. One way to research competing solutions, if you are developing a product, is to search the U.S. Patent Office database for current patents. You should also check the database if you are incorporating a previously developed technological component or process in your product. It is likely that someone holds a patent on at least a part of the process or design that you plan to use. If you use a patented product or idea without the owner's consent, then you could be subject to litigation. See Chapter 8 for a discussion of the nature of intellectual property and for references to databases. For other competing solutions, a good place to start is the Internet. While no one may have developed the exact product you have in mind, there may be many competing solutions for components of the product or for similar products with a different application.

### What Is New and Different about Our Idea?

Ideas are plentiful! Good ideas are rare. When you have an idea for a new product or service, you should ask yourself: Has anyone already thought of this (the answer is almost always yes) and if so, what is new and different about this idea that makes it a good idea at this point in time? Often, the best ideas are old ideas whose time has come.

▼

## GOOD IDEAS WHOSE TIME EVENTUALLY CAME

A personal computer was not a good new-product idea in 1963, when computers cost $250,000+ (in 1963 dollars) and applications were primitive, even though, by then, science fiction writers and futurists had envisioned the personal computer. Fifteen years later, with the advent of the microprocessor, low cost semiconductor memory, high level programming languages, etc., the personal computer became a good idea whose time had come.

The FAX machine existed for decades prior to its explosive market success in the early 1980s. What made the time right were a set of technological advances and the establishment of interoperability standards that made FAX machines useful and affordable for small businesses and individual departments within larger businesses.

▲

Every business case should soberly address the question: Why hasn't this been done before; and why is the timing right now? Good answers would include: a significant increase in the cost of existing alternatives, evolutionary technology advances that make this new product or service feasible now (but not feasible before), new regulatory requirements (for example, requirements to reduce the use of toxic materials) that create a new unmet market need, and a technological breakthrough.

### Why Do We Think We Can Make It Work?

This question is complementary to the previous question (What is new and different about our idea?), but the emphasis here is on the word *we*. There is clearly an unmet market need for an automobile that achieves 200 miles per gallon of gasoline, that is safe and reliable, can achieve highway speeds, and that is roomy enough for a family of five. The problem is that no one knows how to make such a car, and wishful thinking — "When we get the start-up money for our new company, we will figure out how to do it" — is not very persuasive.

There are good reasons to believe that you can make a new product or service idea work even if it is not a new idea. These reasons are similar to those given above for explaining why the timing might be right to bring an old idea to life. They include: evolutionary advances in technology that make the new product or service technically and economically feasible, a technological breakthrough that makes possible the new product or service, and access to very low-cost manufacturing capabilities.

Every business case must include compelling evidence that the technological, economic, legal, regulatory, and other business obstacles associated with the proposed new product or service have been identified and understood and can be overcome. For example, you could make a less expensive and lighter car by removing the bumpers, the airbags, and the seatbelts. Even though this design would satisfy the demand for less expensive and lighter cars, you still must explain how you will overcome the fact that it is not legal to sell such a car in most places in the world and that it is an irresponsible design decision. See Chapter 2 for a discussion of engineering ethics and social issues.

Once you have determined that a design is feasible, you still must determine whether you have the resources to achieve your goals. For instance, given enough time and resources, you could certainly design an operating system and applications that could compete with any of Microsoft's current products. However, you would certainly require more time than is allocated for your design project, and probably more manpower and software engineering experience than you have.

You are limited not only in time and experience, but also in the equipment and software that you will have access to and the money available for components, supplies, or even travel. You must carefully consider each of these and ensure that you have enough of these resources to reach your final design goal. These assets are addressed in different ways and should be considered even before you decide on your final topic. For the formal written and oral presentations of the project, you should create a detailed project schedule, which includes the responsibilities of each team member, shows how much time is allocated to each subtask, and lists who is responsible for acquiring the skills required for completing different tasks. Additionally, you must create a budget that includes lists of equipment, software, facilities, and supplies that you will need. However, even before you are required to produce your schedule and budget, as you are deciding on your final topic, start thinking seriously about what will be required to complete your design. One of the most common mistakes teams make is to assume that because they have a great idea things will naturally work out in the end.

## How Will We Reach Our Customers?

While it is typically not necessary to develop a business plan for your design project, you must identify your customers and consider determining how you will reach them, particularly if you plan to start a business from your project. In order for your new product or service to succeed, you have to make its existence known to prospective customers, you have to be able to accept orders from actual customers, and you have to deliver the product or service to those customers when they place orders. You also have to be able to bill your customers for products and services that have been delivered and you have to be able to collect their payments. In most cases you have to provide customer support services, like answering their questions about problems they are having using the product or service.

Some products and services are sold to a small number of large (in terms of dollars spent) customers. For example, there are a small number of prospective customers for airplane jet engines, but General Electric, Pratt & Whitney, and Rolls Royce are able to remain in business with only those few prospective customers, because each sale is measured in millions of dollars. Success in this type of market is dependent upon one's ability to understand and meet the specific needs and priorities of each of these prospective customers, while controlling costs.

Some products and services are sold to large numbers of small (in terms of dollars spent) customers. For example, consumer appliances are usually in the $10 to $250 range with small profit margins, but the number of prospective customers is huge. Profitability in this market depends heavily on maintaining low manufacturing, distribution, and sales costs.

In each case, there is a value chain. For example, if you want to manufacture a new type of consumer audio component, you could try to sell it directly to consumers by advertising in audio magazines and accepting orders from consumers via telephone, e-mail, and the Internet. Alternatively, you could consider a retailer with existing outlets, such as Radio Shack, to be your (wholesale) customer and have them sell it to (retail) consumers under their brand name, using their stores and catalog sales channels.

Selling large numbers of inexpensive products directly to consumers requires very different expertise and methods than selling small numbers of expensive products or services to large business customers. Every business plan must identify the customers and include one or more credible alternative plans for how the customers will be reached. A credible plan is one that demonstrates that the people who wrote the plan understand who their customers are and that they understand and have planned for the business issues (required alliances, up-front investments, ongoing costs of sales and service) associated with that plan.

The economics of a product (whether you will be able to sell it at a price acceptable to your customers) are part of determining the feasibility of a project. See Chapter 5 for a more thorough discussion of how to build a budget sheet and evaluate the economic feasibility of a product. Chapter 9 contains an overview of the process required for starting your own business.

### What Are the Known Risks?

Every project or business venture (and every design project) has known, anticipated risks. Some of these anticipated risks can be quantified in advance in terms of their potential impact on your project or business venture. These are the "known unknowns" ("unks"). For example, you know that it is difficult to accurately predict the time and effort required to develop a new software application. That is, you know that you do not know very accurately how much time and money it will take. Therefore you identify this as an unknown, and you build contingencies into your project schedule; that is, you consider the impact on your baseline schedule of the software development taking perhaps twice as long and costing, for example, twice as much as your baseline estimates.

Every business venture also has risks that cannot even be anticipated — unknown unknowns ("unk-unks"). For example, you may not even realize that you do not know the complexities and delays associated with selling high-tech products and services outside the U.S. For example, you may have never even heard of export licenses, so it would be impossible to incorporate the time and cost of obtaining them into your project plan. More applicable to novice engineers, you also may not realize how long it will take for you to acquire a new skill. For instance, you may be an expert software developer but not have the extensive experience in hardware design required to complete the hardware portion of a project.

It is difficult to build in contingency plans for unk-unks. However, every project plan should contain a discussion of the potential impact of every unknown that you can anticipate and what your contingency plans are to deal with those unknowns. Contingency plans are discussed further in Chapter 5.

### 1.2.4 Generating and Evaluating Design Alternatives

Once you have determined that your topic is feasible, have formed your team, and have found at least one advisor, you are ready to dive into the project. You will probably be excited about starting and will want to jump in with both feet, but to make the project a success you first have to consider how you and your team will manage the myriad of tasks that will lead to the final design.

At the same time you are developing a project management strategy you should be developing design specifications, since the tasks involved and the project schedule will have to reflect the goals of your design.

As you continue to think about your project, you will realize that you will need to make many decisions about the design: the shape, the structure, the components to develop or purchase, the subprocesses, and so forth. You should perform a formal decision-making or alternative-evaluation process for each of these decisions, even if the decision is easy (for example, if you need a memory chip and are able to obtain a free sample from Company X, while you would have to pay $20 for a chip from Company Y). As you evaluate each alternative you will find that some will not get you to your original design specifications; these alternatives will have to be changed. Also, you may find that a required component is too expensive to purchase so you will have to design your own, or you may perform a stress analysis of a design and realize that an alternative is not feasible so you will have to go back to the drawing board. In each of these cases you should revisit your design specifications and project schedule, updating these as you go. Note that we are not encouraging constant modification of the original specifications or project idea, and we believe that if you do a good job of initially developing a schedule, few changes will have to be made to it. But since you can never predict the unk-unks, redesigns and schedule changes are often necessary. As you complete this project, you will obtain experience in project management, scheduling, and alternative evaluation, and you will start to gain the skills necessary to keep projects running smoothly with minimal design iterations.

In the following we present synopses of major processes that you will be performing: project management and scheduling, developing design specifications, evaluating design alternatives, documenting the design process, and presenting your work. Note that some of these are complex enough to warrant individual chapters.

## Project Management and Scheduling

If you have not had a previous job or project for which you had to develop a project plan, then the idea of formal project management and scheduling is likely to be daunting. However, if you take the time to work on them, both the management and scheduling tasks will go fairly smoothly and will help you to ensure that all components of your design are completed on time.

As you start to refine your topic, you should think about the tasks that will be involved in completing the design, and you should start discussing among the team members who will be responsible for each task. Create a group schedule and make sure that everyone (including your advisor) has a copy.

One of the greatest challenges that many groups encounter is handling group dynamics. By working together on a schedule, all team members can see and, more importantly, agree upon the assigned tasks. Often, however, there are disturbances in the group. Either personalities conflict or team members are not able to complete their required tasks for whatever reason. You should first try to politely resolve these problems in the team, but keep your advisor aware of what is happening. If the problems cannot be resolved by the team members alone, then your advisor can step in to help you work on them.

Chapters 3 and 4 provide details about how to build a cohesive team, plan your project, and schedule all of the tasks. You can use these chapters as guides, but you should also rely on your advisor's experience in carrying through projects to help you judge appropriate tasks and schedules.

## Developing Design Specifications

During the process of narrowing down and evaluating your topic, you informally develop design specifications. For instance, if you thought about the coffee-cup insulator, you might have informally thought of goals such as cheap, reliable, able to fit around both a large and small cup of coffee, easily manufactured, and protects against a typical temperature of hot coffee. As you start to develop your project management plan and schedule, you must articulate these goals, translate them into formal specifications, and evaluate the design alternatives.

Let's take as an example an electronic consumer product called the "Widget." As you were thinking about the Widget as a solution to a problem, you might have started with specifications such as small, light, and easy to use. Your next step is to clarify and put numbers on these specifications and expand with additional specifications. This is one of the places where advisors can lend experience. They often know which specifications can easily be met or might be acceptable to a market segment.

The example below shows some broad design specifications. These specifications include cost, size, input, output, and basic operating modes. These were developed from the original informal specifications by brainstorming about the operation of the Widget and what constraints should be on the operation. If you have a project supplied by a faculty member, industry, or anyone else, then you should consult with them to determine what the needs of the customer are and how the operation of the final project is envisioned. If you are given a set of specifications, you should make sure you can meet them and also consider each to see if you can exceed the goals set by the customer. These initial specifications are overall design goals determined during the feasibility evaluation at the start of the project. As the project continues, the design specifications will expand. For instance, one component of the Widget may be a memory chip. In this case there are multiple ways in which the chip can communicate with a controller. The choice of communication medium is one design decision that must be made, and the communication rate must be included in the design specification. No matter what type of project you are doing, you should repeatedly revisit your specifications to make sure that you will be able to meet them and that you have not forgotten any.

▼

### Sample Initial Design Specifications for a Simple Electronic Consumer Project

| | |
|---|---|
| Maximum Size: | 4" × 7" × 2" |
| Maximum Weight: | 16 oz. |
| Power: | One 9-V battery |
| Operation: | Two modes:  Mode 1 = standby |
| | Mode 2 = full operation |
| Minimum Battery Life: | 80 hours (Mode 1) |
| | 24 hours (Mode 2) |
| User Input: | On/off switch |
| | Fast/slow switch |
| Maximum Speed: | 1600 rpm |
| Minimum Speed: | 800 rpm |
| Output Port: | Infrared output meeting Infrared Data Association (IRDA) standard specifications |
| Maximum Cost: | $7 components |
| | $10 manufacturing |

▲

In the case of a software engineering project the specifications may include topics such as the compatible applications or platforms, the maximum memory required, the speed of operation, and even the layout or order of appearance of graphical user interface screens. Consider the sample initial design specifications given below for the *Coreware* project. This project might be viewed as a software project since the deliverable is Very High Speed Integrated Circuit Hardware Description Language (VHDL). Notice that the specifications are few and straightforward. The deliverable and one operating constraint are specified, as is the system input and output. The remaining specifications are given by a reference to industry standards. Additional constraints could be the complexity or cost of manufacturing an application-specific integrated circuit.

------------------------▼------------------------

### SAMPLE INITIAL DESIGN SPECIFICATIONS
### FOR THE *COREWARE* PROJECT

| | |
|---|---|
| Deliverable: | VHDL description of network level device |
| Maximum Address Translation Latency: | 15 clock cycles |
| System Input: | IPv4 (IPv6) packet |
| System Output: | IPv6 (IPv4) packet |
| Operation: | Meets industry (Internet protocol) standards |

------------------------▲------------------------

## Evaluating Design Alternatives

As you have learned through your classes or your co-op jobs and internships, engineering involves studying systems and evaluating trade-offs. Any successful design process is going to require numerous alternative or trade-off evaluations. You will have to choose between components, performance metrics, procedures, subprocesses, materials, intermediate goals, and a host of other topics and ideas. Two primary techniques, defining the criteria for success and developing decision matrices, can help you in the evaluation of anything that needs to be evaluated. As you decide between alternatives, keep exhaustive notes about the criteria and the evaluation method so that you can look back later and recall how and why you made your decisions.

## Defining the Criteria for Success

For a project to be considered successful, you clearly must meet the required specifications and deliver the project on time and within budget. However, you must determine and give values to the other parameters by which you will measure the results of the completed project. In this step you will define a problem and create an evaluation matrix including weighting factors. To define the problem with sufficient precision for good solutions, you need to define how you are going to evaluate the success of the design — in quantifiable terms. For projects as complex as a building, an embedded computer system, the extension of a rail line, the remediation of a multiply afflicted site, or the design of an integrated circuit chip, there are literally hundreds of parameters that must be within predefined values. Since it is improbable that you will achieve optimality for all of the parameters, it is up to you rank their priority so that you may choose among alternatives during the design process. Once you have the evaluation criteria clearly ranked, it will then be easier to choose between competing alternatives and ensure that the decisions are unbiased. One method of summarizing the good aspects and drawbacks of each alternative is to develop a decision matrix.

### Developing Decision Matrices

If you have a small number of criteria or predetermined values by which you can compare your alternatives, then it is easy to put these in graphical form in a decision matrix and view the alternatives side by side. Assuming that you have chosen and ranked the decision criteria, you first assign a weighting factor (possibly 1) to each criterion. Next, select a scale by which to evaluate each alternative. Scales can be ranges of numbers (1 to 10), ranges of values (poor to excellent), or even colors (light to dark red) or graphics (empty to completely filled circles). Finally, objectively rate each of your alternatives, put ratings in a matrix, weight each rating, and combine the entries to produce one value for each of the rows.

As an example, let's consider purchasing a memory chip for the Widget. Possible evaluation criteria are the cost, size, and type of input/output lines. We will assume that we are looking at chips with two possible connectors and that we prefer one slightly over the other. Obviously there can be many more criteria, such as reliability, shipping time, or access speed, but for this example, we will use only these three. Let's assume that the size is not very important to us, but we are on a limited budget (and we can obtain free samples of the chip from Company A, while Company B would charge $10 and Company C would charge $20), so cost is an important factor and thus deserves a high rating. Therefore, we will pick a scale of 1 to 5, where 1 is poor and 5 is excellent, and assign the weights of 3, 1, and 2 for cost, size, and access format, respectively. We can now construct the decision matrix as illustrated below. Note that we have weighted each of the rankings to determine the total in the right-hand column. It is now easy to see that Company A's chip is the winner, but that Company B's chip is a close runner-up. If the criteria or alternatives change, or if we decide to add new criteria to try to broaden the difference between the top two choices, it is easy for us to go back, add another column or row or adjust weights, and reevaluate the alternatives.

▼

#### SAMPLE DECISION MATRIX

| Alternative | Cost (× 3) | Size (× 1) | Format (× 2) | Total |
|---|---|---|---|---|
| Company A | 5 | 1 | 3 | 22 |
| Company B | 3 | 1 | 5 | 20 |
| Company C | 2 | 5 | 3 | 17 |

▲

Note that we have emphasized unbiased or objective rating. Often teams start the decision-making process with a predetermined goal; that is, they know which alternative they want to win. Then they consciously or unconsciously select the criteria, weights, and ratings so that their favorite comes out ahead. This is poor engineering practice and leads to bad design decisions. You should fairly evaluate each alternative and agree to accept the outcome, no matter what it is.

## 1.2.5    Simulation and Modeling

Modeling and simulation of systems are integral parts of the design process. Imagine building a satellite system. Even if you have an enormous budget (millions or billions of dollars), you cannot afford to make mistakes. Once the satellites are in orbit, there is almost no chance to retrieve them and make design corrections. During and even before the initial design phases, many subsystem models (for example, what is the action of the solar panels unfolding?) need to be built and many system simulations (for example, how will the satellite rotate to face the earth?) need to be performed.

Even if you are working on a project with much lower risk, and you have budgeted time for corrections and design modifications, modeling and simulating your system or components can be a useful part of design verification and can save a great deal of time during the implementing and testing phases. Modeling can be as high-level as creating a block diagram of your system, with associated simulation being pencil-and-paper views of flows. However, you likely know from classes that you can do software modeling and simulation of circuits using either free academic or commercial packages. We do not advocate any particular package but rather suggest asking advisors for suggestions on appropriate products.

## 1.2.6    Implementing Your Design

Implementing your design should naturally fall out of all of your previous work. You have determined what needs to be implemented, in what order the components should be built and assembled, and who is responsible for each part. Your goal at this stage is to implement as rapidly as possible, testing at each implementation stage (see Section 1.2.7 for more information on testing). However, even if you have divided the implementation into stages, allocated resources to each stage, scheduled time for development, and acquired all necessary components, that does not mean that your implementation will go smoothly. This is the time when the unk-unks are revealed. Common pitfalls teams encounter are misread spec sheets, lack of the proper tools or equipment for implementation, lack of experience with the needed tools, incompatible software versions, and underestimating time required.

While your goal in engineering design is to deliver a completed design meeting customer specifications, with prototype or final product, at the specified time, invariably you will eventually participate in a project that falls behind schedule, typically due to poor planning. What are you to do when you fall behind schedule or find that you need a partial or complete redesign? As will be discussed in Chapter 5, you have several options at this point. The first option (assuming you can acquire needed components in time) is to work a heroic number of hours to rework the project and possibly even recruit help from friends and colleagues. The advantage of this is that you can have the project done on time and no one will be the wiser (except for the people who have to pay for new components). The disadvantages are clear. You lose nights and weekends to the project and you risk not having the project completed on time (a new design will have unk-unks too). Your second option is to discuss the schedule slippages with your manager or advisor. The advantages of this are twofold. First, someone with more design experience can often suggest ways to speed up the project. Second, you have prepared your manager or advisor for the worst. They will not be surprised if the project is not completed on time and can help you either scale back the project, negotiate new specifications with the customer, or at least let the customer know to delay other engineering projects that are dependent on yours (thereby saving the customer significant idle time). Since your annual review or class grade depends on the level of completion of the project, you will still have to put in long hours to make up for lost time and you may still receive a low grade or poor evaluation, but now your customer, advisor, or manager realizes that you are someone who will do well on the next project.

### 1.2.7 Testing and Verifying the Design

For novice engineers, testing and verification of the design is typically an afterthought. Teams put it off, hoping there will be sufficient time at the end of the project to do evaluation. However, to have an effective design, you must develop a test plan when you start the project and test throughout the design process. No matter whether you are designing a new machine, a software system, a chemical process, or a new bridge, you should test the design components (controllers, power supplies, software subroutines, etc.) individually and then test each subsystem as it is implemented. Finally, you must test the system as a whole. The component and subsystem tests are obvious; you need to test to make sure they work as the manufacturers claim or as your design specifies. The final system should not only be tested to make sure that it works as specified by the customer, but should also be tested for safety and for usability (whether someone unfamiliar with the product can easily understand how to use it and all of its features). It is also a good idea after initial (alpha) testing by the designers to have colleagues, peers, and customer representatives, that is, people who are unfamiliar with the product, do beta testing. These people can likely capture bugs that designers cannot, since they do not know how the system should work. If you cannot find people who have the time or would be willing to test an entire system, you also have the option of giving these external testers guidelines for what operations or features you would like tested. For instance, if you are concerned about the insertion operation for a database or the sequence of switches needed for a given system state, then specify that your user should concentrate on these operations. One of the easiest methods of determining how to test a design is to develop a formal test plan. Initially, a test plan sounds like a great deal of work for something that you could probably easily do informally. However, it is a good reminder during the project of what you have tested and what remains to be evaluated.

### *Test Plans and Procedures*

Test plans or test procedures can range from a broad overview of what should be tested to a book-sized document that details every test move to be made. The testing phase begins as your project begins, since at the outset of your project, you should have at least a general idea of what you must test (you must meet customer specifications). As stated above, if you are building a tangible consumer product, you should not only test to make sure that each component works independently as specified and that the integrated system functions as expected, but also test to make sure that it is user-friendly as well as durable and safe. If you are building a software application, you should test for its usability, portability, fault tolerance, and security. If you are designing a bridge, you obviously have to test for strength and durability under the expected loading and also put in a sufficient factor for safety.

The example below shows the preliminary outline for testing *The Talking Book* (see the draft report in the Appendix for details on the design) and its components. This project is mainly a hardware project requiring some code for a microcontroller. The test outline is high-level; it can serve as a reminder of what is to be tested. It is a document that should be initially written as the preliminary design is being discussed, but should also evolve with the project. As the designers change components, add more features, or even think of additional tests, the new tests should be listed and performed. Note the progression of the tests. The designers first test components individually and then test interoperability. Finally, they test the entire system for specifications and then perform endurance and user tests. The designers start from this simple test outline and determine when to test each component and subsystem. For instance, each component should be tested as soon as it is purchased.

What the designers learned from doing component testing early is that the anticollision properties of the reader were not as robust as desired. The team went back to the drawing board, designed a check for multiple tags in the field, and determined that they had to wait until three tag readings were received. Another design constraint that became apparent was the large current draw, causing the designers to rethink their power supply. With each test new constraints arose. These are the types of issues that are not always clear from manufacturer specification sheets, particularly when you are adopting technology to a product for which it is not designed. Thus, by testing early, designers are able to determine potential problems and modify their initial design as they progress through the project.

————————————▼————————————

### SECTION OF A SAMPLE TEST OUTLINE
### FOR *THE TALKING BOOK*

RFID tags
> Test reader output in presence of each tag (check output pin and company specifications).
> Sequentially add tags dispersed in space to determine anticollision limit.
> Move tags in and out of range to simulate pages turning; note maximum and minimum distance of tags from readers and speed of detection.
> Check current draw on reader.

ISD ChipCorder
> Test all input and output pins.
> Test storage limits.
> Test ability to write to and read from device.
>> Check memory addressing.
>> Check output power.
>> Check memory permanence and memory erasing.
> Determine current draw.

> .
> .
> .

Test RS-232 cable.
Test switches.

Test speaker.
Test microphone.
Test microcontroller code.
    Mimic RFID reader input to microcontroller; monitor output lines.
    .
    .
    .

Test system.
    Write and play back pages (sequentially, nonsequentially, step
        through all combinations).
    Test maximum write.
    Test speed of switches.
    Determine maximum speed of system (number of page flips/second).
    Battery life testing.
    Usability testing (give to new user with no instructions).

————————————————▲————————————————

Obviously, test plans and procedures will vary based on the type of project you have selected. You may be testing both hardware and software, or you may place much more emphasis on usability and human interface issues if you are designing a software system or product meant for wide commercial distribution. Therefore, you should work with your advisor, manager, or mentors to ensure that you develop an adequate plan for testing your project. Unless you are required to publish the test plan or results separately, you should include the detailed test plan and procedure in an appendix of all project proposals and reports. In either case you must report honestly on the outcomes of the tests.

No matter how formal a test procedure you follow, it is vital that you start testing and verification as soon as possible; that is, test your components independently as soon as you receive them or formally and extensively test each graphical user interface piece or software routine as soon as it is developed. If you start your tests early during the design process, you can then modify the design. However, if you wait until your prototype is complete to see whether all of the components work together and that it meets specifications, you will likely not have enough time to make any necessary modifications in case that it does not.

### 1.2.8 Documenting and Presenting Your Work

During the design process, you will most likely be required to develop at least a project proposal and a final report. Often you will be required to produce interim (status) reports, give oral presentations, develop formal test plans, and produce user documentation. Although it may often seem like busy work at the time, documenting your decisions, procedures, and results, and then later presenting these, are important parts of a design process. Documenting is the process of recording design decisions and work performed; this is typically for your own benefit. Presenting your work is the process of conveying your ideas or accomplishments to others. They are two separate processes, but if you do a thorough job of documenting during your design process, then presentation of your work should be fairly easy.

## Documenting

Clear documentation of your entire design process will enable you to organize your thoughts and will save time when you make design decisions and when you present your work. As was noted previously, even while you start the initial brainstorming for topics, you should record all ideas, no matter how farfetched they sound at the time. Often teams want to go back to ideas. Having a written record can help you recall how you reached an idea and will let you branch off from previous ideas, rather than starting by repeating the entire thought process leading up to a topic.

Similarly, you should keep a written record (either a paper or electronic version) of all design decisions and of all work performed. This includes a list of all resources used, all Web pages visited, all patents or books read, and all people consulted. Very often you will want to go back to material referenced. Having the entire reference on hand, whether it is a URL or a name and phone number, makes it easy to do so. Each design trade-off should be thoroughly explained in writing. If, during the course of your project, you need to revise your design, then you can go back to your documented alternatives and select the next best alternative, without going through an entire reevaluation process. Clearly list and detail the work responsibilities of each group member. This list can serve as a contract among the group members and will help remind team members of the tasks they have agreed to perform.

While all of the team members should be responsible for documenting their work, the team can also assign a recorder (secretary), either on a permanent or a rotating basis. The recorder will record the minutes of the meetings and keep track of all of the documentation. One of the traditional methods of documenting individual work is to have each team member keep an engineering notebook that is dedicated to the project. By keeping a bound notebook with an accompanying folder for printouts or loose papers, the project is easily organized. Each activity related to the design project should have a dated entry in the notebook. Then it is easy to go back and see the sequence of events that led to a design decision, and it is easy to find a record of the design alternatives examined. The group recorder should keep a paper or electronic notebook for the entire group, as well as an individual notebook. In addition to serving as a reference for pulling together presentations and revisiting design ideas, a dated notebook can justify ownership of intellectual property. It is very important that the notebook be a bound book so that pages cannot be removed or inserted. Each page should be dated and signed at the bottom by the individual that made the entry. If the work could possibly lead to a patent, then each page should be witnessed. Loose pages should be described in the bound book, dated, and preserved. See Chapter 8 for more information on patents and intellectual property.

## Presenting

If you have been thorough in recording your design decisions, alternatives, evaluation criteria, motivation, and background material, then all you have to do to present your work is to weave the documented process into a coherent written or oral report. Whenever you do a project, whether for a course or in industry, it is necessary to present your work for a number of reasons. First, at the start of a project, you would like approval of the project from management or from an advisor. Then you must also convince people to support your project, either financially or by other means. As the project develops, you present your work to others to obtain their feedback on the process or on your design decisions. Finally, after your project is complete, you present your work to make it known to those who are evaluating your performance or who will be interested in using your design.

Since the goals of each of the presentations vary, and the audiences may vary as well, you will have to tailor each presentation to the particular outcome you desire. For instance, if you are asking an organization for a grant to support your project, you should provide a high-level overview of the project that is understandable to those outside your field. However, if you are presenting your work to a group of faculty that will be grading you on it or a group of colleagues who are experts in the area, you should present some motivation for your project, but also present details about the engineering design decisions and test results. Clearly, tailoring the presentation to your audience is a lot of work, and therefore you should start early on each presentation and have it reviewed to make sure you are actually conveying the information you wish to give your audience.

Chapters 6 and 7 provide suggestions for how to present your work in both written and oral forms. However, one of the best ways to learn how to present your work is to look at what others have done in the past, and to get a feel for how they have dealt with arranging different topics and handling the different presentation tones for various audiences. Sample reports and slides are given in the Appendix. These documents are excerpts of reports and presentations from two senior design projects and are representative selections of how projects can be presented. While you may not necessarily want to follow the style of these reports, they are available to give you an idea of how to start the writing and presenting processes.

## 1.3 CONCLUSIONS

A design project will require that you use in a coordinated and creative manner appropriate components of the knowledge you have acquired during your engineering education. All engineers should hearken back to skills, knowledge, or experience from classroom situations, jobs, hobbies, or general reading that may pertain to the design effort.

You are asked first to define a problem, generate a range of solutions, evaluate those solutions using economic and engineering criteria, and develop a viable solution. The deliverables are not just plans and specifications, but a detailed feasibility study and preliminary design and prototype, in which constraints have been accommodated, conflicts resolved, and costs and benefits reasonably well defined. Further, decisions will have been made about layout, size, shape, materials, processes, methods of construction, and so forth. This product should be completed to the stage at which a client could — on the basis of your report — decide to build (or operate) or not and could go on to write a complete contract for the final design and construction (or operation).

Next you are called upon to present your ideas in written, graphic, and oral form in as professional and convincing a manner as possible. To do so you will have to work in teams, set your own schedule (within that required by your university or company), decide which of the many resources and tools available to you are appropriate, and prepare a series of both oral and written presentations of your work.

How is a design project different from course work? Why should this not be an opportunity to vacation for a month or so and then work hard for the duration of the project? The answer lies in your freedom of choice. In prior projects, you were told the requirements — what to include and what not to include, as well as most of the detailed steps en route. As you complete your course work design project or are given your first industry design project, you are in charge. Your first challenge will be to decide on a project, or, more properly, a problem to be solved (within the scope of projects acceptable to your university or needed by your company). It must be interesting enough to hold your attention, yet it must not be so complex that you will never accomplish anything. Making an accurate survey of the work required, setting the end point where it belongs, writing a clear and plausible proposal, and estimating the design cost realistically separate firms that prosper from firms that fail.

You — not your advisor — are responsible for defining the criteria by which your design success will be evaluated at the time of the final presentation, and this sort of definition is a demanding process. You must then decide what work must be accomplished, divide it equably among the team members, and schedule yourselves over a prolonged period to complete an enormous variety of tasks. Finally, you are expected to carry the analysis and presentation to a far greater level of detail and polish than ever before.

## FURTHER READING

Dym, C.L. and Little, P. *Engineering Design: A Project-Based Introduction.* New York: Wiley, 2000.

A short introduction into general engineering design based on several examples oriented toward civil and architectural engineering. In particular, the discussions of interacting with clients and developing design alternatives are useful.

Pahl, G. and Beitz, W. *Engineering Design: A Systematic Approach.* London: The Design Council, 1988.

A highly systematic, precise approach to design, particularly tailored for mechanical engineers. It is nearly overwhelming in its detail but potentially useful for those whose approach is extremely step-by-step.

Salt, J.E. and Rothery, R. *Design for Electrical and Computer Engineers.* New York: Wiley, 2002.

This book gives a solid overview of the design process from an electrical engineering point of view. It includes useful case studies with detailed discussion of how the different components of a project are reflected in these cases.

Sullivan, W.G., Lee, P.-M., et al. A survey of engineering design literature: methodology, education, economics, and management aspects. *Engineering Economist,* 40, 1994.

This paper provides a comprehensive list of relevant books and articles pertaining to the subject area of engineering design, with specific focus on the methodology, education, economics, and management aspects of the design discipline. The purpose of this survey of engineering design literature is to provide a compendium of books and articles that may be instrumental in facilitating the process of sharing and disseminating design knowledge.

Voland, G. *Engineering by Design.* Reading, MA: Addison Wesley, 1999.

A large textbook written primarily from the mechanical engineering point of view following the standard design approach: formulation, teamwork, creativity, decision making, communication. It contains many examples and interesting case histories. It particularly emphasizes ethical issues and hazards analysis and a variety of issues specific to product design such as manufacturability, reliability, and recyclability.

★ *Key Points*
*Chapter 2*

- Since engineers have a profound effect on the environment and on the well being of the planet, their actions matter.
- There exist many authoritative resources for engineers who face ethical questions. Among them are engineering codes of ethics and texts that discuss methods of conflict resolution and analysis.
- There exist several key questions that engineers should ask themselves before undertaking new projects. Among them are:
  - Who is going to be affected?
  - What is the effect of the project on natural resources?
  - Is development of the product safe?
  - Is development of the product ethical?
  - What is the effect of the project on human welfare and on human rights?
  - What could go wrong?
  - Can this product be used unlawfully or unethically?
  - What is the potential risk? What is the potential liability?

# Chapter 2

# ETHICS AND THE SOCIAL IMPACTS OF DESIGN PROJECTS

*Moshe Kam*

## 2.1 WHY ETHICS?

### 2.1.1 The Blood-Carrying Pipe

When I was a student at Tel Aviv University in the 1970s, one of my instructors used to tell the following story. When the instructor was a student, he had to answer a long question in an undergraduate exam on fluid mechanics. The question was about a pipe carrying blood from the Mediterranean Sea to the Dead Sea. Exam-takers were provided with a large number of technical parameters of blood, heights and profiles of the terrain between the two seas, tables, and many other necessary and unnecessary details. What was apparently missing was an explanation of why would anyone engage in the bizarre and totally absurd endeavor of transporting blood between two large bodies of water, which are scores of kilometers apart. Interestingly enough, not one of the more than hundred test-takers saw fit to comment on this small issue. The students all merely plunged into the calculations, did their best to find flow rates, masses, and pressures, but ignored completely the ridiculousness of the underlying task.

### 2.1.2 The Plagiarism Detector

A colleague of mine, a young professor who came to visit my department at Drexel University, recently presented a seminar on "a new educational tool," which he proposed for routine use by all faculty members. It is a very sophisticated plagiarism detector, able to detect unattributed paragraphs lifted by students from Internet sources. Not only is the tool able to search the World Wide Web for copied sentences and paragraphs, but it is also able to recognize when these have been slightly (or even heavily, sometimes) edited and paraphrased to mask their original source. The software can further flag sharp changes in writing styles within a submitted essay, which is a telltale sign of plagiarism, even if the exact source of copied material cannot be located. Finally, the new tool keeps track of a "writing sophistication index" for each student in the database, developed on the basis of past submissions. It raises an alarm when the index of new submitted homework is significantly higher than the student's long-term average.

It was clear that my colleague was very proud of his product. The algorithms pushed the state of the art in computer linguistics, database search techniques, automata theory, and discrete event systems. The accuracy and speed of the machine were unparalleled (he used parallel computing and simultaneous coordinated searches). It was a truly multidisciplinary project at the cutting edge of computer science, computer engineering, and linguistics. My colleague had data that demonstrated the efficiency and efficacy of his design, and comparison tables showed how he beat all rival designs of this kind, at least those whose work was reported in the open literature. Most attendees were impressed, and there was a sense of appreciation and enthusiasm when the prepared remarks were concluded.

Then the question-and-answer period came.

My colleague was very surprised when some of the attendees expressed serious — and at times angry — reservations that had little to do with his new algorithms or the sophisticated search techniques. Rather, they objected to the very idea of building and using such machines in a university (or other public environments). Serious doubts were expressed about the advisability of screening all work of undergraduate students at the university through my colleague's wonder machine. Are we not subjecting all of our students to the equivalent of a compulsory search, assuming that they are all presumptive plagiarizers? Does this kind of inquisitive analysis conform to the ideals and the objectives of an academic institution? Of a free society? Does this kind of presumption of guilt constitute a good example for our students to follow when they make decisions about privacy and professional treatment of colleagues and subordinates in their own professional careers? My colleague was completely unprepared for the small storm that erupted. He and the seminar moderator were trying — in vain — to keep the discussion technical, focused on the science and on the performance graphs. A small yet vocal group in the audience kept them from this objective, hammering at the ethicality of the design and the design philosophy. Some of the critics did not fail to remind the embarrassed speaker that he was standing in downtown Philadelphia, just a few blocks from the Liberty Bell and Independence Hall. The moderator concluded the seminar a bit early, somewhat hastily, in an atmosphere of growing contention and discomfort. It was clear that the proud inventor was totally focused on the technology but paid very little regard to the social implications of his invention.

### 2.1.3 Fighting Crime in Cyberspace

My colleague is not alone. The search for illicit uses of the Internet is the focus of vibrant — and necessary — inquiry by thousands of scientists and engineers, most of them employed by governmental or quasi-governmental organizations. They are not looking for college students too lazy to read "Death of a Salesman," but rather for real malevolent terrorists, big-time thieves, traders of illicit narcotics, and distributors of child pornography. Yet when you hear some of them describing their work, serious questions arise about balancing the threat with human rights and professional norms, and the presumption of innocence, which is a hallmark of Western judicial thought,* becomes an important issue, as it did in my colleague's unfortunate seminar. Here is a quote from an interview with one of the better known "cyber warriors":

> Our programs analyze sentences such as "I sent you ten yams and five lemons" and have to decide whether the sender of the message is a greengrocer or a terrorist who is informing someone about a shipment of explosives.... We want to know everything. We want to know who's using the Internet and how they are using it. "Who's who in the zoo" is the best description I can offer of our motivation: we want to know where everyone is located, in which cage. If he changes his color, like a chameleon, and disappears, we still want to locate him using our method of operation. We want to identify transfers of money, knowledge or instructions of terrorist bodies. (Dror, 2004)

Implicit in this goal description is the assumption that we are allowed to read the electronic mail of an innocent greengrocer, one who has indeed sent his client ten yams and five lemons, with little regard to the greengrocer's rights. There is also an implicit assumption that a utilitarian principle — we are trying to defend society from terrorists — allows us to ignore the ethical principles known as "respect for persons" — for example, the privacy and property rights of the greengrocer. In the interview with the "cyber warrior," he acknowledges that his "programs are liable to infringe on the privacy of hundreds of millions of people who have nothing to do with terrorism." However, he "is not losing any sleep over this."

In the context of cyber counterterrorism, the goal of engineering ethics is to make the sophisticated engineers and scientists, armed with their supercomputers and government grants, lose a little sleep over the privacy rights of the greengrocer and his or her client, and by extension, of all people in our society.

## 2.2 SOCIAL IMPLICATIONS

Engineering matters. Our designs, plans, blueprints, and manuscripts are translated into buildings, roads, bridges, communication networks, radars, sensors, actuators, control systems, and vehicles — a plethora of structures, devices, tools, and machines that control and change the environment. The impact of these on society is not always clear at the time that they are made or invented. Vladimir Zworykin, the inventor of television, could not have foreseen *Seinfeld*. The father of information theory, Claude Shannon, probably did not have in mind anything like the compact discs and DVDs that have spawned from his theoretical 1948 papers in the *Bell System Technical Journal*. As another example, think of the impact of the Benjamin Franklin Bridge, connecting Pennsylvania and New Jersey across the Delaware River. It is certainly a technical marvel, now and at the time of its construction, and yet very little was written — when it was built or since — on its contribution to the deterioration and decline of Camden, NJ, in the twentieth century. Many inventions, designs, and plans undertaken by engineers do have obvious societal implications. Often, these are not subtle, hard to predict, or too distant in the future to be foreseen. Engineers who build and erect a 150-ft tower for cellular telephony in the midst of a pristine natural forest know, even before the design commences,

---

* "The maxim, innocent until proven guilty, has had a good run in the twentieth century. The United Nations incorporated the principle in its Declaration of Human Rights in 1948 under article eleven, section one. The maxim also found a place in the European Convention for the Protection of Human Rights in 1953 [as article 6, section 2] and was incorporated into the United Nations International Covenant on Civil and Political Rights [as article 14, section 2]" (Pennington, 2001).

that their actions will have negative environmental consequences. These consequences need to be assessed against the benefit of providing access to the emergency 911 service in the forest. Engineers who build tracking devices for convicted criminals at the request of law enforcement authorities, or design listening and monitoring devices, can and should understand that these devices can be used on innocent civilians such as the designers, not only on convicted criminals.

## 2.3  QUESTIONS AND TESTS

In analyzing the societal and ethical implications of design projects, several questions and tests are worth reviewing. They can help identify these implications, assess their severity, and sometimes offer solutions or ways to mitigate difficulties. This is not to say that all moral and ethical dilemmas have a satisfactory solution. Nor can we promise that thorough reading of books on ethics will provide guaranteed algorithms to resolve such dilemmas. While often a creative middle way can be found between conflicting principles and objectives, some cases have no such solution. Some proposed projects and ideas are better not implemented, because their potential to do harm exceeds by far their potential to do good. Some guidance on the general assessment of projects and actions is given in Chapters 2 and 3 of Harris et al. (2005).

### 2.3.1  Ask Yourself: Who Is Affected?

Quite often the moral dimension of a design or plan does not become clear until all groups and individuals affected are considered. Think not only about the people who have cellular telephones (whom you want to help) but also about those who do not (but may be affected by your invention or design). Think about non-Americans (the raw materials that your project needs may have to be supplied by them). Think about nonhumans and even inanimate objects (animals, plants, the environment in general, the skyline of Philadelphia if your invention would alter it). Assess the impact that your project would have on all these groups, individuals, and objects, including all affected nonusers. Apply the classical Utilitarian and Respect for Persons tests (Harris et al., 2005, Chapter 4) to determine whether the proposed actions are ethically permissible.

▼

Proponents of drug testing in third-world countries often hail the benefits that new experimental drugs can bring to the individuals who participate in these tests. Moreover, the participants are often compensated financially. These testing proponents often underemphasize the fact that test participants are deprived of the benefits of drugs and treatments that are already available in industrial nations. Moreover, often the developed drugs are never available in the third-world country where they were tested, because they are too expensive there.

▲

### 2.3.2  Ask Yourself: What Is the Effect of the Project on Natural Resources?

In this category you need to explore the raw materials that the project would consume, their availability, the long-term impact of their use on the environment, the waste generated by production, and alternatives (especially renewable resources). Include resources consumed or affected in the course of normal future use of your product and consider product disposition and decommission. Design for recycling is always preferred, but you should also think of encouraging actual recycling, not just enabling it (most recyclable devices and products are never actually recycled). Key questions are whether the resources could be used in a more effective or economical manner and whether we should hold some natural resources in reserve for future generations. Consult Chapter 9 in Harris et al. (2005) if any of these issues becomes significant.

---▼---

Recently the New Jersey Pinelands Commission has struggled with the need to comply with the Telecommunications Act of 1996.

The New Jersey Pinelands are the U.S.'s first National Reserve — an important ecological region consisting of 1.1 million acres of forest with aquifers containing 17 trillion gallons of pure water.

The Telecommunications Act of 1996 gives providers of cellular telephony services who hold valid FCC licenses certain rights concerning the erection of antenna towers. The basic rationale was to provide society with wide access to mobile telephony services, especially in emergencies (for example, 911 services).

Allowing high antenna towers in the Pinelands would have detrimental effects on the reserve, increase industrialization and pollution, and mar the landscape. Is there an acceptable balance between the benefit of cellular telephony and the benefits of the reserve? Should we sacrifice some of these benefits (for example, being able to use our cell phones when we drive to Atlantic City) in favor of others (the pristine beauty of the pygmy pine forest)?

---▲---

### 2.3.3 Ask Yourself: Is Development of the Product Safe?

Many design projects require experiments on humans and animals. Projects that have requirements of this kind need to comply with institutional and legal requirements that would guarantee conformity with standards of experimentation that are accepted by the research community and the law. These include informed consent in the case of humans. You may need to look beyond these standards to ascertain that the potential benefit warrants the experimentation.

Prototypes that involve moving parts, vehicles, flying objects, high voltage, and use of chemicals have the potential (sometimes the grave potential) to cause harm. Part of the ethical checklist is to ensure that the risks involved were assessed and reduced to an acceptable level.

---▼---

A few years ago a senior design group in my university developed a breathalyzer that slowed and eventually stopped a car when it became clear that its driver was legally intoxicated. This was a very sophisticated device that had a high probability of detection and a low probability of false alarm. However, the designers forgot to assess the effect of the device on the safety of other drivers on the highway, and further analysis showed that a car slowed by their mechanism is likely to cause car accidents no less serious than the ones they tried to prevent.

---▲---

### 2.3.4 Ask Yourself: Is the Development of the Product Ethical?

Development of new products often involves competition with existing ones. It always requires money and other resources. It is sometimes tempting to cut corners by using existing patents without permission, infringing on trade secrets by reverse engineering, using illegally copied software, or asking for free samples of components and devices under false pretenses. Members of design teams and their advisors need to be sensitive to these potential infringements, as they tend to become habitual and condition their perpetrators to a lifetime of deceitful practice.

### 2.3.5 Ask Yourself: What Is the Effect of the Project on Human Welfare and on Human Rights?

Does the project affect or restrict freedom of movement, freedom of speech, freedom from unlawful search and seizure? Many past design projects have involved devices that monitor the location or activities of individuals (on the road, online), and legitimate questions were raised about potential misuse or deliberate misapplication. This is a prime area where safeguards and proper monitoring of use of the future product may be offered. Products that pose significant potential to infringe on human rights may have to be overhauled or even abandoned (recall our example above, Fighting Crime in Cyberspace [Section 2.1.3]). Quite often, "rights tests" (Harris et al., 2005, Section 4.6, p. 89) are applicable.

### 2.3.6 Ask Yourself: What Could Go Wrong?

What is the potential that the outcome of your project will contribute to accidents and harm (to users or bystanders)? Is the product safe? How can it be made safer? Is it safe for children and disabled persons? Is it easy or likely that the product would be used in an unsafe way (even if this occurs in spite of warnings and written instructions)? Can you predict some of the unintended consequences?

▼

The authors of fire, electrical, and building codes often promote new revisions of existing codes in the name of increased safety. The revisions are often meant to reduce the hazard from fires, unsafe constructions, etc. At the same time, many of the new codes were proven too costly to incorporate in older neighborhoods, with negative effects on downtown areas. Developers shun away from modernization efforts in these areas, and many blocks become ghost towns, hotbeds of crime and poverty. This example demonstrates the unintended consequences of attempts to do good, and the need to predict as many of these consequences and account for them (in this case, by exempting some areas and some buildings from some of the new revisions).

▲

### 2.3.7 Ask Yourself: Can This Product Be Used Unlawfully or Unethically?

I recently heard an interview with the authors of a book about corporations and individuals who "cheat America" by not paying taxes. The authors have complained that their book, written in order to raise public outrage over this kind of cheating, is now used by some to learn and execute sophisticated tax-evasion techniques. Does your invention or device have the same deficiency? Is your product likely to find use among criminals or the morally unscrupulous? If it does, in what ways can this effect be reduced? Is the risk of unlawful use tolerable, considering the potential benefit that will accrue under lawful and ethical use of the device?

### 2.3.8 Ask Yourself Questions about Risk and Liability

What is the risk to the consumer involved with the use of your product? What is your potential liability (as well as the liability of a manufacturer who will produce large quantities)? How would potential customers be made aware of the risk? How does the risk compare to the present situation in the market?

## 2.4 ACTING ETHICALLY

In addition to reviewing the questions and tests as you progress through your design project, there are a number of strategies you can take to ensure that you work in an ethical manner.

### 2.4.1 Understand Your Relationship with the Sponsor

All industrial projects are governed by the policies of the company, and the relationships between companies and their customers are carefully negotiated by the parties involved. You must be aware of these relationships. Similarly, coursework projects might involve an industrial or an academic sponsor; nevertheless, all of these projects fall under your university's intellectual property guidelines. The best defense against misunderstandings over owner-ship and rights is to review thoroughly and meticulously, at the beginning of the project, all the details of all policies, agreements, contracts, and memoranda of understandings between all organizations and all individuals involved in the project. Many of the issues that Harris et al. (2005) covered under "Engi-neers as Employees" (Chapter 8) are applicable. Be aware of the supremacy of written university policies over most other arrangements and understandings and insist that all matters concerning intellectual property be managed accord-ing to this policy, and in writing. Be aware of the complications and conflicts that can ensue if you allow your right to publish to be impeded.

▼

In 2001 there was a legal dispute between the University of California at San Francisco and Immune Response Corporation of Carlsbad, Cal-ifornia. The company demanded damages after university researchers published data indicating that one of the company's drugs was not effective against HIV. This is one example of many potential disputes between the sponsor of research and the researchers, and it demon-strates the potential conflict between business interests and the public welfare. In this dispute the university prevailed, but we do not know how many similar cases ended differently, with the researchers sup-pressing data that could benefit the public because of contractual rea-sons or expectations for future funding.

▲

### 2.4.2 Keep Records

Undocumented, messy, and disorganized research and development (R&D) is increasingly recognized as unethical R&D. When activities are not summarized in writing, dated, and explained, major efforts can be derailed. Work cannot continue in the next design cycle due to lack of documentation. Disputes over ownership, time of invention, and scope of effort flare.

As discussed in Chapters 1 and 6, you should keep all your work together, put all project-related comments in a well-organized and dated notebook, and insist on similar behavior by all your teammates.

▼

One of the most contentious scientific misconduct disputes of the 1990s involved a paper published in 1986 in the journal *Cell* by Thereza Imanishi-Kari, David Baltimore, and several collaborators. The dispute, known now as "the Baltimore Affair," involved congressional hearings, op-ed articles in the *New York Times*, various actions of law enforcement authorities (including analysis of laboratory logs by the Secret Service), and multiple accusations of fraud, data cooking, misinformation, and cover-up. One of the sources of confusion and dispute was the highly disorganized records of the original experiments. It is widely believed that much of the aggravation suffered by all parties to this affair (lasting more than a decade) would not have occurred if clear, clean, marked, and dated records of the experiments were created and kept.

▲

### 2.4.3 Finally, Speak Up

If ethical and societal-impact issues present themselves in the course of the project, raise them in project meetings and insist on resolution. One of the most common mistakes is to assume that these matters are secondary, may somehow solve themselves later, or would fall in someone else's domain. Often these assumptions are proven wrong, and seemingly secondary matters stop otherwise sound technical ideas from being realized. The time to speak up about ethical and societal questions is always at the time of their appearance — not when ignoring them is not possible any longer.

### REFERENCES

Dror, Y. Cyberspace Warrior, an interview with Abraham Kandel, Haaretz, April 15, 2004.

Harris, C.E., Pritchard M.S., and Rabins, M. (2005). *Engineering Ethics: Concepts and Cases*, 3rd ed., Belmont, CA: Wadsworth.

Online ethics center for engineering and science (includes links to codes of ethics, and case descriptions), http://onlineethics.org/ (accessed July 17, 2004).

Pennington, K. (2001). Innocent until proven guilty: the origins of a legal maxim. In *A Ennio Cortese,* 3 vols. Rome: Il Cigno Galileo Galilei Edizioni. Available at http://faculty.cua.edu/pennington/ (accessed April 15, 2004).

Hekert, J.R. (2000). *Social Ethical and Policy Implications of Engineering.* New York: IEEE Press.

A recent collection of articles, including good discussions of sustainable development, technology and health care, and information technology.

http://www.ieee.org/portal/index.jsp?pageID=corp_level1&path=about/whatis&file=code.xml&xsl=generic.xsl (accessed July 19, 2004).

The IEEE Code of Ethics, which can be found from a link on the home page of the IEEE, www.ieee.org.

Johnson, S.F., Gostelow, J.P., and King, W.J. (1999). *Engineering and Society.* Englewood Cliffs, NJ: Prentice Hall.

This text provides a wide overview of the societal impact of engineering decisions and analyzes engineering efforts from historical, sociological, and economic perspectives.

Strand, M. and Golden, K.C. (1997). Consulting scientist and engineer liability: a survey of relevant law. *Science and Engineering Ethics,* 3, 357–394.

This is a well-written summary of the various doctrines and principles that guide the American legal system when faced with legal actions concerning the work of engineers.

★ *Key Points*
*Chapter 3*

- Project management can save time and improve quality.
- You must be clear on your project goals.
- Remember that you will need other resources: people, money, and machines.
- Scheduling can be considered at multiple levels: team members can devote only a few moments to quickly develop napkin lists, or one person can be assigned a full-time job as a scheduler.
- The greatest benefits come from the least complex level, but don't neglect considering the more complex levels.
- The essence of effective scheduling consists of defining tasks, their logical relationship, the necessary resources, and the time required to complete the task.
- Keep the number of tasks to a reasonable level.
- You can effectively make use of your computer to aid in developing schedules and managing your project.

# Chapter 3

# PROJECT MANAGEMENT

*James E. Mitchell*

## 3.1   INTRODUCTION

### 3.1.1   What It Is and Why You Want It

Why not start designing right away? You have an idea, some friends, and a lot of time. Let's get it done quickly and take some time off, or get the patent application approved. How complex can it be?

Some groups may be lucky. They may have worked together before or they may have a strong, experienced leader who has run projects and will intuitively follow the right process. Most groups, however, are not that lucky. They spend considerable time making false starts and working far from their potential. The result is often a design that either is not as fully developed as possible or takes more time than necessary.

Project management, a fully developed profession,* focuses on achieving the best result through the most effective use of all resources. Learning the project process and using the appropriate tools can save you considerable time and improve what you create.

### 3.1.2   You Do It Already

Of course you are already engaged in project management. Going to the store for food or undertaking a homework assignment are both projects; each is a "temporary endeavor undertaken to achieve a particular aim" (Project Management Institute, 2000). You do not consciously plan for

---

* The best location for a good overview of the profession is at the Web site of the Project Management Institute (PMI®). This chapter is structured using their concepts, although the emphasis and details are adapted to the design experience in college.

them or engage in resource management or any of the five process groups (initiating, planning, executing, controlling, and closing) or nine knowledge areas (project integration management, project scope management, project time management, project cost management, project quality management, project human resource management, project communications management, project risk management, and project procurement management) defined by the Project Management Institute (Project Management Institute, 2000). Nonetheless, by remembering to take your wallet or deciding on the grade you want for the homework, you are in fact undertaking project management.

### 3.1.3 Formal Project Management

What distinguishes the need for formal project management is the scope or critical nature of a project. In industry the money involved, a crucial delivery date, or scarce, heavily scheduled equipment can all require active project management, even if only for a brief period. In school the two usual motivations are the length of time involved or the necessity of working in teams. Throughout this chapter we will focus on working within a set time frame with limited resources to achieve specific goals, while recognizing that other issues may become important in special circumstances.

#### *Goals*

The first step in managing a project is defining the aim or goal. From a college senior's point of view the aim is often as unformed as achieving another grade or fulfilling the last graduation requirement (the advisor is usually focused on something different — what you learn). The first major task of a design project is usually to convert these goals to a clear set of specifications related to the desired characteristics of the completed design. The sooner the team is clear on these goals, the more effective they will usually be in making progress through the endless range of distracting possibilities toward a successful result.

#### *Time Frame*

For a college group the time frame is usually established by the senior year course requirements, so little specific effort is needed to establish the overall deadlines. Each year, however, a number of groups start early to expand the time available. In industry the company management or the client defines the time frame.

Within these overall limits, however, the group has considerable leeway in how to achieve their goals. The scheduling section of this chapter is devoted to an exploration of how defining the necessary tasks, their logical sequence, and the people responsible for them can assist you in meeting the deadlines.

#### *Resources*

- **People:** In all projects, people are your most important resources. Teamwork is a critical issue in any big project. We have devoted Chapter 4 to this issue.
- **Money:** Chapter 5 addresses the subject of money in detail. Planning for the acquisition of "real money" (sponsors, your own pockets) and when it will be needed will save you grief during the year; for example, big copying bills can be a very unpleasant surprise.

- **Hardware and Software:** For certain projects you must use a particular machine or piece of software — usually an expensive, fully scheduled piece like an electron microscope, a chip fabricator, or a testing machine. Planning the process and talking to the individual responsible so that the machine will be available when you need it can make a big difference. Once that commitment is made you may need to shape much of your design process around that need. In that case the importance of scheduling increases dramatically.

## 3.2 SCHEDULING: THE KEY SUPPORT TOOL

### 3.2.1 Scheduling for the Working Professional

Every product, whether the B-1 bomber, an electronic circuit, a highway bridge, a vaccine factory, or even a university degree, depends on a schedule. Without a schedule the product's purchasers do not know when they can use it. Without a schedule the manager cannot hire the workers to produce the project, order the ingredients, or provide the necessary physical environment. Without a schedule for final exams a student will not graduate.

When you first graduate you will initially be working with schedules set by others. As you advance in responsibility, however, you will be setting them yourself, especially if you have ambitions to start your own company or become a manager. Your success will be judged in significant part by how well you follow the schedule you set — come in on target and you are a star; come in too far ahead and you will be accused of protecting yourself too much. Miss your deadline and you become a goat.

The specifics expected of scheduling vary from industry to industry. In a fast-paced world missing a deadline by even a few months can be the difference between a company thriving or collapsing into bankruptcy. In the slower world of large-scale military development schedules are almost expected to slip — though there always has to be a justification. In both worlds schedules are vital and are constantly reviewed and revised.

### 3.2.2 Scheduling for the Student or Early Professional

Unless it was required in a class, you have probably never bothered with developing a project schedule in school. Completing a homework assignment or a term paper is straightforward — you know the steps and move through them, usually with a late night or two at the end. Most students look on a formal schedule as something approaching "make-work," an unnecessary addition to an overloaded life. For most school projects you are right. It is easy to organize in your head, perhaps with the aid of a calendar on which key due dates are recorded.

A major design, whether in school or in industry, should be different. You are embarking on a months-long project with multiple teammates. By necessity, each team member will probably undertake 30 or more separate tasks. That makes well more than a hundred separate tasks to be completed for a typical team — two or three hundred would not be uncommon. Many of those tasks should not be begun until others have been completed. There is a logic to completing them efficiently and fully that is almost certainly too complex to be understood completely in your head.

If that is so, then why did your friend or brother or aunt go through college without taking a schedule seriously? Most likely they were able to do so by:

- working more hours than necessary or
- leaving out some desirable ingredients of the design process or
- working to a lower quality standard than possible.

The most effective way to avoid one of these outcomes is to prepare some level of formal schedule and to communicate this schedule to others to obtain feedback. In this chapter we will present different possible levels, the types of schedules, and the tools to work effectively, with comments on the amount of time and effort required. There is a good chance that you will have a more enjoyable senior year if you use these tools.

## 3.3 THE BASICS OF SCHEDULING

Scheduling is simple in principle, and often in practice. Only six elements (deadlines, tasks, logic, effort, resources, and costs) are necessary to build a schedule for any project. By using them one can plan anything from dying a piece of cloth to building a Mars probe. They are presented here in decreasing order of importance for the type of projects for which you are likely to be responsible in college or early in your career, that is, you are almost certain to need deadlines but are unlikely to want to spend the necessary effort on or be responsible for predicting and tracking full project costs.

### 3.3.1 Deadlines

You know what a deadline is: an absolute date by which something must be complete and ready to be handed to someone else. In college the project deadlines are presented by the department coordinators for each term — the dates specific documents must be turned in, presentation dates, and so on. In industry the client or your manager, or both, sets that deadline.

### 3.3.2 Tasks

A task is a piece of work, something that takes time, whether it is a minute to make a decision or 5 years to carry out a piece of research. Its key characteristics are that it has duration, that one or more people (or sometimes machines) must do something, and that at the end there is a product or outcome. Usually there are preceding tasks that must be completed before a particular task can be begun. In more complex scheduling one can accommodate partial completion of tasks.

Any given project can be defined in terms of a set of tasks that must be completed individually before the entire project is completed. Thinking through these tasks is probably the most useful effort for a college or early-professional team. Even if you do not take the next step of organizing them temporally, you will have a better idea of what must be done to complete the work of the project.

Defining those tasks at a useful level of detail is also one of the most difficult aspects of developing a realistic schedule. If you take a simplistic approach, they could be as few and simple as "define project," "develop ideas," "make presentation," and "write report." The problem with this level of detail is that you probably have not learned anything helpful. You have to work with more specific tasks to really learn anything (for example, "define input parameters," "define evaluation criteria," "test first order solution"). Experience with many college design projects suggests that 15 to 40 major tasks and subtasks per term is likely to be a helpful number, but still comprehensible and manageable.

**41**

CHAPTER 3
Project
Management

### 3.3.3 Logic: The Network

Once you know the tasks (or, in reality, in parallel with developing them) you can determine the logical relationship between them. Normally this is just a matter of deciding which tasks follow each other in sequence. Often many tasks can be in progress at the same time (usually undertaken by different team members to balance work loads), but inevitably there are decision tasks for which multiple prior tasks must be complete before a decision can be reached.

Preparing this logic network can be extremely productive for any design team, even if the next step of determining specific dates is not taken. By agreeing on the sequence of tasks it becomes easy to assign them to individuals so that they can all be productive throughout the entire life of the project, not just at one time. There are often straightforward tasks such as testing a component or preparing a format for a presentation that can be assigned and accomplished while waiting for another major task to be completed.

Discussing the network logic with your advisor or manager can be particularly beneficial. Since they have more experience with projects, they may see tasks that are left out or improperly sequenced. Even if they approve the entire plan, it is useful to inform them of what you intend to do through the duration of the project.

Particularly with some software scheduling programs, it is possible to get a graphic representation of the logical network. This is usually presented as a series of boxes representing tasks, with lines between them representing logical sequence requirements (Figure 3.1). Time is almost always represented as flowing from left to right. For those familiar with computer programming, this network is essentially the same as the flow chart for the development of a piece of software — noting that what would be an iterative loop in a program must be accomplished by a sequence of evaluation and repeat tasks since time is linear.

### 3.3.4 Effort: Time to Complete Tasks

As with the logic network, once the tasks are determined, you can decide how long each should take to complete. By assigning a length of time to a task you can build a calendar that reflects the dates when the tasks should be under way or complete. The most usual of these calendars is the Gannt chart (named after Henry Gannt, an early twentieth century researcher on efficient work), a horizontal bar chart in which each line represents a task, the beginning of the bar, the start date of the task, and the length of the bar, the time required to complete it.

It should be evident that if you have a logic network for your tasks and the length of time that each takes, it is an automatic process to generate the Gannt chart. Once you have assigned the start date for the project you or, more usually, a dedicated computer program can work your way through the diagram calculating when the tasks must occur and when the project will be complete. An example of a network, explored in greater detail in Section 3.6, is the following set of tasks with the logical relationships between them.

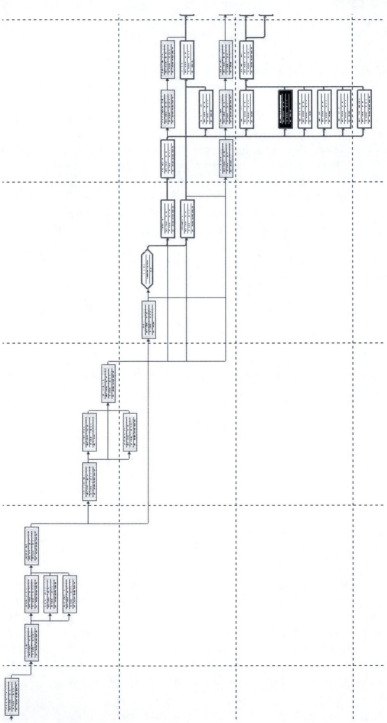

**FIGURE 3.1**
Sample scheduling logic
network.

When there are parallel tasks in a logic network, it may mean that some tasks have "float" time, because those of shorter duration do not need to be started at exactly the same time as a longer-duration parallel task in order to have them completed at the same time. The computer programs that perform scheduling all show the float time as well as related concepts such as "early start" (the earliest time a task can start) and "late start," the latest time that a task can start without delaying completion of a project.

Another computation that flows from assigning time to tasks is the ability to identify the "critical path" through the network. This is the sequence of tasks that must be completed exactly on time to complete the project without delay. This concept is so important that many people know formal scheduling as "critical path method" (CPM) scheduling. Special shading or color on the network logic or Gannt chart often identify the critical path.

## Duration vs. Working Hours Effort

One refinement of assigning time to complete tasks is to distinguish between the actual working hours necessary to complete the task vs. the number of calendar days necessary to complete it. Working in actual hours of effort allows one to assign multiple resources to a task (when feasible) and thus to shorten it, whereas the calendar length of time does not take the possibility of extra assigned resources into account. Where possible the actual hours approach is usually more beneficial but usually takes more effort to determine initially.

## Common Time Estimation Errors

Experience is by far the best guide for estimating how long it will likely take to complete a given task — experience that most new design team members do not have. One of the great benefits of college design experience may be that it allows you the opportunity to make mistakes on time estimation when the consequences are relatively low. The following suggestions are offered to help avoid the worst mistakes. Also, see the sample schedule for a capstone design project discussed in Section 3.6 for estimates of time required for selected tasks or see sample project schedules from previous design project reports for hints on task times.

- **Design should take time.** Good design is almost always an iterative process involving incubation, first ideas, analysis, revision, expansion, more iteration, and then documentation.
- **Outside individuals rarely respond rapidly**. It may take several weeks to get an answer from a manufacturer or code official.
- **Writing takes time**. To write a good technical document you need to allow time to organize it, write it, obtain a review, revise, then print and copy the report. Two weeks is often short for a really good job.
- **Preparing a good presentation takes time.** As with writing a technical report, you need to organize the material, prepare the graphics, create the presentation, rehearse, revise, and rehearse again.

## 3.3.5 Resources

When you assign a task to a team member you are assigning a resource to that task. For most college design projects the team members are the only resources necessary. In some cases, however, other resources might be necessary and must be scheduled. A good example would be a fabricating or testing machine that is heavily used or expensive and must be scheduled far in advance. For the purpose of your project, that machine is a resource that can be assigned to a task such as "fabricate test chip," perhaps along with one of the team members to operate the machine. You would want to develop your schedule to be sure that all other tasks necessary for that task are complete on time and to arrange that the resource has to be used for only the time available. Making resource assignments is particularly beneficial in determining that all members of a team have something interesting to do and that the amount of effort is equally divided.

As with the network logic, it may be helpful to discuss the resource allocation with your advisor or manager. Many advisors want assurance that each team member is doing real engineering, not just secretarial or machinist work. Resource assignments can aid this discussion and perhaps highlight potential areas of difficulty.

### 3.3.6 Costs

The final ingredient in the scheduling stew is the costs of the various resources. If you assign an hourly cost to each resource (both people and machines, and even fixed costs such as purchases), then you can turn your schedule into a cost estimating and tracking tool. Since college students are working for free, this aspect of scheduling often seems the least important to them. Once you are in industry and working on a large project it will become vital if you are responsible for its budget.

As with the critical path scheduling, the reality of attempting to keep track of costs is complex enough that it really does not become practical until you start using a dedicated tool. Almost all contemporary scheduling programs build such capabilities in so you can produce reports of project costs just as easily as scheduling with a Gannt chart.

## 3.4  HOW TO SCHEDULE

### 3.4.1  Levels of Scheduling

You can devote little or much time to scheduling. The relationship between time invested and the reward for doing so is one of diminishing returns — the greatest rewards usually come from the initial work, with less and less return as you invest more time. It is for that reason that very detailed scheduling is worthwhile only on very large projects (practically, in the millions of dollars) when the budget allows for one or more people devoted to managing a schedule. In school or during your first industry assignment you are likely not working at that level, so you may want to pick one of the lower levels discussed below.

### *Simplest Schedule: An Organized List*

While entering a few dates on a calendar could undoubtedly be called scheduling, the simplest formal effort would be an ordered list of tasks that attempts to be comprehensive for the project. At least some of them would have dates attached. If it is a short one-person project, this list is often mental — like the four stops you decide to make before going to the gym at 3:00. We all do this kind of scheduling every day and never think of it as a formal schedule, yet it is vital to keeping our lives organized.

With more than one person involved, it is a simple step to put this list on a piece of paper or in an e-mail to be shared by the participants. With this exchange the dangers of misunderstanding the process of completing the project are greatly reduced. If names (resources) are attached to the various tasks, then responsibility and equity of effort can be addressed early in the project rather than the too-often-seen recriminations of slacking that advisors or managers must deal with late in a project's life.

This level of scheduling can usually be completed in less than half an hour on a multi-person project or in a few seconds for a simple day's list. It certainly has the highest benefit-to-cost ratio of any scheduling work undertaken and necessarily precedes any more extensive work.

## Medium Level Schedule: A Gannt Chart

Time is not visible. It is one-way and very slippery, yet it must be managed to successfully complete large projects. One of the simplest tools for making time visible and to some extent manageable is the Gannt chart. Gannt's contribution was to realize that instead of a series of boxes on a calendar, time could become one of the axes of a bar chart — usually the horizontal axis. When the start and stop dates of the tasks (one task per line) are plotted, one has a visual representation of what will happen when working on a project. It is quick to create (a spreadsheet package such as Microsoft's Excel is a natural) and can reveal a great deal about the nature of a project.

Vertical lines indicate what should be the progress on each task at any date. Overlapping bars indicate areas of potentially heavy demands on personnel and thus times when slippage may occur. Deadline dates can be readily indicated so time-to-deadline is visually evident.

Preparing a Gannt chart depends, of course, on having a complete list of the project tasks and length of time estimates. It is not strictly necessary to have worked out the complete logical relationships between the tasks, but the work of placing tasks reasonably on the chart is greatly helped by first creating a logical network.

For a college design project a Gannt chart of 20 or 30 tasks can be thrown together in less than an hour using a software package such as Excel (since some revisions are almost always necessary a computer is highly beneficial). Taking another half hour to fine-tune the relationships by thinking through the logical relationships either before or afterwards will make the chart far more useful. The reality in college capstone design courses has very often been that a Gannt chart is created the day before the presentation and adjusted to show how well the team met their schedule. This looks good on the screen but has not actually done the group any good during the term. The whole point of this chapter is to convince you that the time spent planning at the beginning will pay off during the project.

## Complex Schedule: A Scheduling Program

It should not be surprising that scheduling has drawn the attention of computer programmers. Its economic importance as well as its fundamentally numeric and graphic nature makes it an obvious candidate for using software to increase efficiency. There is good range of available programs (some discussed in Section 3.5.3) with costs ranging from nothing to hundreds of thousands of dollars, offering capabilities ranging from simple list-making to "enterprise" tracking of all projects for a multinational corporation. Almost all operate by keeping the fundamental elements defined above (deadlines, tasks, logic, effort, resources, and costs) in an internal database, then displaying computations using those elements either in graphical or report form. The primary differences between the programs lie in their differing abilities to deal with large numbers of tasks and in the range of reports and charts they can produce. The simplest can be learned in a few minutes, whereas the most complex require weeks or more of training and constant use for proficiency. Here we will talk about the level of functionality likely to be useful to a college or early-career professional project. You can undoubtedly learn about the more complex programs through an employer or from the manufacturers' Web sites.

All but the simplest of the programs ask that you provide them with at the least the deadlines, tasks, and effort required, allowing you to enter further information as you wish. Many allow creation of subprojects to work with a subset of the overall number of tasks that can be considered independently of the larger project. The more complex programs take into account the limited nature of resources (including cash in some cases) to spread out the work realistically. They allow you to look at the same information in many different formats and levels of detail, all in the service of managing the project efficiently.

### Using a Complex Program for Your Design Project

If you have a very powerful program available, the temptation may be to use as many of its features as possible. If you have ambitions to become a project manager or executive, then taking the opportunity in a first major design project to learn the process where the penalties for failure are low may be worthwhile. For most first-time project managers it is more realistic to use only a lesser program or to limit your use of a bigger program. With one of the intermediate programs (such as Microsoft® Project*) you can probably expect to spend 3 to 5 hours learning how to use it and then the same again preparing your first schedule. An update will probably require 1 to 2 hours if no major changes are required. The following suggestions are given with this limited approach in mind:

- Learn the program by following the tutorial that comes with it. That tutorial will almost always give you the best instruction on the specifics of the program's implementation of the principles described above.
- Limit the total number of tasks for your first project to fewer than 100 — more will probably overwhelm you.
- Wherever possible use actual hours for each task rather than the number of days. You will need to adjust the calendar to a reasonable number of weekly hours, but that will make real estimating of effort much easier.
- Make each team member a resource. Assign each task to one member.
- Use cost capabilities only for overall project costs — usually by assigning hour rates to the resources. More detailed costing will probably bog you down.
- Be wary of attempting weekly updates — the time required is probably not worth it. One or two updates during a project phase can be very worthwhile, especially when reviewed with your supervisor.

---

*Microsoft® is a registered trademark of Microsoft Corporation in the United States and other countries. Screen shots appearing in this chapter are reprinted with permission from Microsoft Corporation.

## 3.5 TOOLS FOR SCHEDULING

The previous sections described the elements of scheduling and the various levels. Here we will describe the tools available. The approach reiterates some information presented before but should allow ready selection of the appropriate tool for a team's goals (see Table 3.1).

### 3.5.1 Calendar: Paper or Computer

Do not forget the simple wall calendar, pocket organizer, or computer calendar. If all your important dates (exams, labs, trips, holidays, and so on) are visible simultaneously it will make planning your work realistic. Using a paper list for group work can be difficult but is still sometimes beneficial.

### 3.5.2 Paper and Pencil

Simple lists, initials, and comments are often the beginning of a much more comprehensive scheduling effort. There is still nothing to beat a pencil and paper (or blackboard for a group) in terms of rapidity and flexibility of displaying relationships, rough ideas, or equations simultaneously. The difficulty comes when one wishes to revise or present such information to others. Then translation to a word processor or other program becomes very useful.

### 3.5.3 Electronic Tools

The Web, as is no surprise, is the most rapidly developing area for scheduling. Because almost all projects worth scheduling are team efforts the Web is a natural medium for sharing. The individual software programs are almost all able to create Web pages showing their information, while the more complex ones allow direct Web interaction. Below we highlight one simple Web-based software package that can be extremely useful for design teams; however, it is possible to create your own Web site to manage your team's schedule and documents. Similarly, spreadsheet packages can aid in developing schedules. Aside from other uses in a design project, spreadsheets are extremely useful for several aspects of the scheduling process: making lists of tasks, sorting the lists, extracting subsets of information, and developing Gannt charts. Beyond this, there are many specialized tools for developing and tracking schedules. The detailed schedule in Section 3.6 makes use of Microsoft Project.

### *eProject.com: An Example of a Web-Based Project Tool*

eProject Express offers a fee-based calendar, task tracking, messaging, and file sharing for its members. It provides almost all the benefits of local computer–based system, but maintained on the Web so it can be readily shared will all team members. One can create tasks assigned to individuals, with due dates and comments about those tasks. They can be viewed, updated, completed, and sorted by a variety of criteria, with automatic e-mailing to the involved individuals. It goes beyond the capabilities of a program like Microsoft Project because it offers collaboration tools (Calendar, Shared Documents, Email, Task Assignment in particular) that can be extremely beneficial.

eProject is one member of a broader category of products that are often called collaboration tools. These tools all extend well beyond the restricted

| Tool | Deadlines | Tasks | Logic | Effort | Resources | Costs | Comment |
|------|-----------|-------|-------|--------|-----------|-------|---------|
| Calendar | Good | Poor | — | — | — | — | Quick. Allows understanding of relationship to individual's entire life. |
| Paper and pencil | Good | OK | Poor | Poor | OK | Poor | Quick and versatile. |
| Web — eProject | Good | Good | OK | OK | — | — | This Web-based service offers various levels of service at increasing fees. The evaluation used here is based on their "workgroup" version. One of its significant benefits is the ability to have a central document repository for easy sharing. |
| Spreadsheet | Good | Good | Poor | Poor | OK | Poor | Like pencil and paper but easy to revise and produces good Gannt chart output. |
| **Specialized Scheduling Programs** | | | | | | | |
| TurboProject LT | Good | Good | OK | Good | Good | — | A free, basic scheduling program intentionally limited in its scope to give you incentive to upgrade to their inexpensive commercial programs. Very good for basic capstone design work. Easy to learn and use. |
| MS Project | Good | Good | Good | Good | Good | Good | The industry standard for many small to medium projects. Can do everything described above. Retail cost is about $400. Takes several hours to learn to use effectively |
| Primavera | Good | Good | Good | Good | Good | Good | Very high-end, complete program (by reputation; the author has not used it). Typically used only for large, complex projects because of the learning and costs involved. |

**TABLE 3.1**
A Comparison of Scheduling Tools

domain of project scheduling addressed in this chapter but can be extremely worthwhile in a large project, particularly one that is geographically dispersed.

### Spreadsheets

One of the key features of spreadsheet software is that it makes it easy to list and sort tasks. In a spreadsheet, such as Excel, the "filter" or "list" commands make this process extraordinarily simple. Exploring their use is highly advisable. The example shown in the detailed project schedule below in Section 3.6 includes some useful categories of information that a team might wish to use in planning a project.

A good spreadsheet program's charting capabilities make preparation of Gannt charts extremely easy. In the simplest form, you put the dates across the top of the chart and tasks down the first column. You can then use the cell "fill color" toolbar to color the cells, perhaps using a color code for the different types of activities or the person performing them. If you are more ambitious it is possible to create a more precise chart. For example, if you choose to use Excel, Microsoft's Web site contains instructions for doing so (Article #213447 in the Knowledge Base).

## 3.6   A SENIOR DESIGN SCHEDULE EXAMPLE

The following schedule illustrates most of the issues raised in the discussion of project management. It should also serve as an example of the benefits of using specialized software, in this case Microsoft Project. The author (Prof. Mitchell) spent about 6 hours learning how to use the software following the tutorial that comes with Microsoft Project and looking through the help files. The actual schedule (based on observation of a dozen years of capstone design projects) took about 4 hours to prepare. It attempts to be useful to all disciplines, although the author's civil and architectural engineering background may have influenced some specific tasks.

The assumptions of this schedule are that four potential members of a group start thinking about their design project after the end of the spring term of their junior year. During the period before the start of the first phase they form the team, agree on the general nature of their project, and solicit an advisor. During the first phase they work up to 12 hours per team member per week to accomplish all the tasks defined — quite a lot of time in the experience of many capstone design teams, but necessary to achieve a first-quality proposal in at least some departments. The next phases would be add-ons to this schedule, following the same principles.

The schedule follows them through the first phase, through their first presentation, and to the delivery of their proposal. It takes advantage of many, but hardly all, of the features of Microsoft Project, including resource allocation, costing, and specialized calendars.

### 3.6.1   Overview of Scheduling Tasks

In what follows, each illustration (Figures 3.2 through 3.12) is a portion of the Microsoft Project screen. Because of the size, the entire screen is not shown, only representative sections. Discussion of each illustration considers it from the point of view of either the process (using Microsoft Project) or the example (the senior design experience).

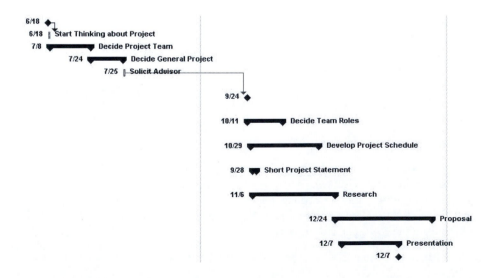

| | | ❶ | Task Name | Duration | Jun '01 | Jul '01 |
|---|---|---|---|---|---|---|
| | 3 | | ⊟ Decide Project Team | 34 hrs | | 7/8 ▼ Decide Projec |
| | 4 | | Hold Initial Meeting | 2 hrs | | 6/19 ▯ Hold Initial Meeting |
| | 5 | | Evaluate Technical Strength | 8 hr | | 6/24 ▨ Evaluate Technical Stren |
| | 6 | | Decide Quality Goals | 8 hr | | 6/28 ▨ Decide Quality Goals |
| | 7 | | Evaluate Interest In Project | 8 hr | | 7/3 ▨ Evaluate Interest I |
| | 8 | | Meet - Agree on Team | 8 hr | | 7/8 ▨ Meet - Agree ( |
| | 9 | | ⊟ Decide General Project | 28 hrs | | 7/24 ▼ De |
| | 10 | | Brainstorm General Ideas | 8 hr | | 7/12 ▨ Brainstorm |
| | 11 | | Evaluate Possible Sponsors | 8 hr | | 7/17 ▨ Evaluate |
| | 12 | | Check Interests of Advisors | 8 hr | | 7/22 ▨ Che |
| | 13 | | Meeting - Agree on Project | 4 hr | | 7/24 ▨ Me |
| | 14 | | Solicit Advisor | 2 hr | | 7/25 ▯ Sc |

**FIGURE 3.2**
Entering the data.

## Process

Entering tasks is like preparing a list in MS Excel. Type a task on each line and enter an estimate of how long it will take. Note that in this example the durations are set in hours. Since Project displays durations in days by default, it was necessary to adjust the *calendar* for the project as well as the display settings. Using hours makes much more sense than days for a capstone design project.

The schedule also takes advantage of Project's ability to form subtasks by indenting a task under a general heading. This allows "collapsing" of a complex set of tasks to get an overview of the entire project.

Project also uses the basic tool of linking tasks to determine which must precede which — accomplished by selecting two tasks and clicking on the Link button. This linking establishes the logic of the project. It, plus the estimates of duration, generates the Gannt chart on the right.

## Example

Figure 3.2 shows the very beginning of the project, starting in June. It emphasizes the importance of forming a compatible team that shares the same quality aspirations and complementary skills.

## Project Summary Gannt

6/18 ◆
6/18 ▮ Start Thinking about Project
7/8 ▼━━━━━━ Decide Project Team
7/24 ▼━━━━━ Decide General Project
7/25 ▮ Solicit Advisor
9/24 ◆
10/11 ▼━━━━━ Decide Team Roles
10/29 ▼━━━━━━ Develop Project Schedule
9/28 ▼ Short Project Statement
11/6 ▼━━━━━━ Research
12/24 ▼━━━━━━━ Proposal
12/7 ▼━━━━ Presentation
12/7 ◆

**FIGURE 3.3**
Project summary
Gannt.

By creating subprojects you make it possible to display a summary of the entire project as in Figure 3.3 without distracting detail. You can quickly move between a complete summary like this or partial ones showing varying levels of detail for individual subprojects.

*Example*

One sees in Figure 3.3 that the work before the start of the first phase does not take very much time, as it should not since you are not officially working on the project yet anyway. Nonetheless, starting this early is wise because it gives you plenty of incubation time and also allows you the highest probability of finding the advisor of your choice.

Note that this view highlights a problem with this particular schedule that might have been lost in the detail: the subproject "Develop Project Schedule" extends for too long. In a revision of this schedule more resources might be devoted to this subproject to shorten the time it takes.

## Network Detail

*Process*

The network view of a project (Figure 3.4) can be used either to enter or review the logic of a project. It does not display time in the way of a Gannt chart but does display most clearly the logical interaction between the various tasks as well as critical information about each task.

That one can switch between two different views (in fact an almost bewildering number of views) of the same information emphasizes that the heart of any project management program is a database. All the views are different ways of looking at selections from that database.

*Example*

That many things can, and indeed should, happen simultaneously is made particularly clear in the network view of the information.

## Network Complex

*Process*

There is little difference between Figure 3.5 and the previous, more detailed network diagram.

*Example*

The network complex illustrates the ability to view at different graphic scales onscreen, thus allowing either focus on individual portions of the network or comprehension of the entire process.

## Resources Table

*Process*

When the resources are entered one can assign hourly rates and overtime rates as well as a specific calendar. These assignments are used in calculating project costs and how long in days a specific task assigned in hours will actually take.

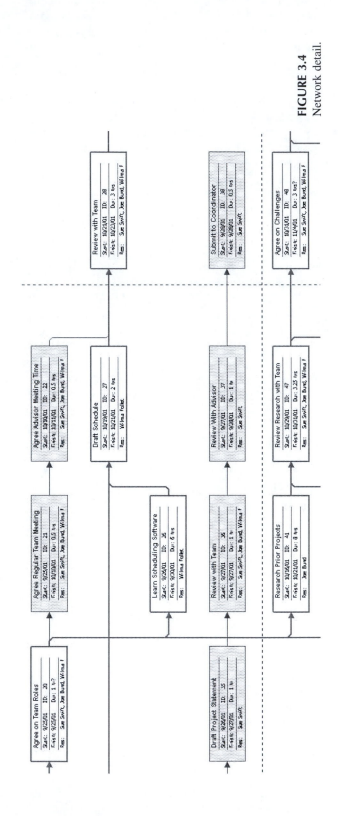

**FIGURE 3.4**
Network detail.

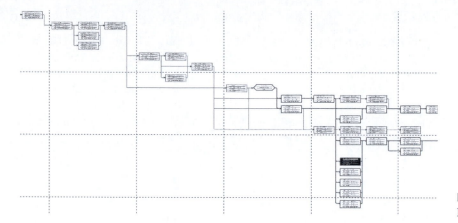

**FIGURE 3.5**
Network complex.

| Resource Name | Initials | Max. Units | Std. Rate | Ovt. Rate | Cost/Use | Accrue At | Base Calendar |
|---|---|---|---|---|---|---|---|
| Sue Swift | SS | 100% | $100.00/hr | $150.00/hr | $0.00 | Prorated | Senior Design |
| Joe Bund | JB | 100% | $100.00/hr | $150.00/hr | $0.00 | Prorated | Senior Design |
| Wilma Follet | WF | 100% | $100.00/hr | $150.00/hr | $0.00 | Prorated | Senior Design |
| Phil Green | PG | 100% | $100.00/hr | $150.00/hr | $0.00 | Prorated | Senior Design |
| Dr. Knowledge | DrK | 100% | $200.00/hr | $500.00/hr | $0.00 | Prorated | Advisor |

**FIGURE 3.6**
Resources table.

Note that the Cost/Use (Figure 3.6) column is not used in this case. If you had a machine that needed to be rented with a lump-sum rental fee, then there would be an entry made in that column.

## Example

The cost rate assumes that overhead is included. It is the billing rate for each individual. The calendar created for this project assigned 2 hours per day for each day except Saturday. The advisor's calendar assigned 1 day per week.

### Resource: Individual Hours

| ⊟ Sue Swift | 189.05 hrs |
|---|---|
| Start Thinking about Project | 2 hrs |
| Evaluate Possible Sponsors | 8 hrs |
| Evaluate Technical Strengths | 8 hrs |
| Decide Quality Goals | 8 hrs |
| Evaluate Interest In Project | 8 hrs |
| Meet – Agree on Team | 8 hrs |
| Solicit Advisor | 2 hrs |
| Fall Presentation Date | 2 hrs |
| Generate Tasks Required | 2 hrs |
| Review with Team | 2 hrs |
| Meeting – Agree on Project | 4 hrs |
| Brainstorm General Ideas | 8 hrs |

**FIGURE 3.7**
Resource:
individual hours.

*Process*

You can create resources for any project. In Figure 3.7, four students as well
as an advisor were identified and assigned a calendar specific to the time they
had available for capstone design (12 hours per week for each student and 1
hour for the advisor). By assigning one or more resources to each task, the
database records the time they are spending. It is then possible to create a
report that displays the list of tasks and how much time each individual spends
on each.

*Example*

In this view of Sue Swift's work many of the tasks are ones in which all
members of the team participate (their hours summary would show the same
hours). Other tasks, such as specific research work, would have differing hours
for each team member.

Note also that Figure 3.7 shows that this draft schedule allocates too many
hours for some tasks. In a revision these would be adjusted to be more realistic.
This over-allocation arose because of an idiosyncrasy of the way MS Project
assigns default hours and then adjusts them when resources are assigned to a
project.

### Resource Graph

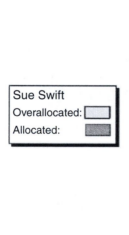

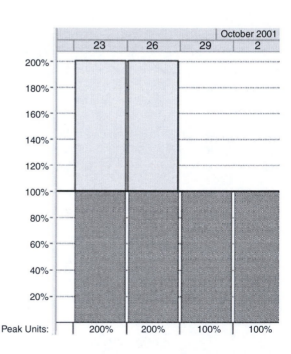

**FIGURE 3.8**
Resource graph.

*Process*

One of the many ways of viewing the data presents the percentage of "possible"
use for any given resource. Reviewing that information can show bottlenecks
in the project, ones where people will need to work overtime.

Most serious project management programs allow a process called
"resource leveling." That process uses an algorithm to stretch out the time
taken by a task in order to reduce individual overloads.

*Example*

**55**

CHAPTER 3
Project
Management

In this case, Sue Swift is scheduled to work twice as much as she has said she is available October 23 through 26. It is possible that the revisions of task time would reduce this overload, or assigning someone else to help her during that period could reduce the load.

## Resource Usage Hours

| Resource Name | Work |
|---|---|
| ⊞ Unassigned | 0 hrs |
| ⊞ Sue Swift | 189.05 hrs |
| ⊞ Joe Bund | 189.05 hrs |
| ⊞ Wilma Follet | 193.3 hrs |
| ⊞ Phil Green | 190.55 hrs |
| ⊞ Dr. Knowledge | 10.2 hrs |

| Details | November | | | | | | December |
|---|---|---|---|---|---|---|---|
| | 10/29 | 11/5 | 11/12 | 11/19 | 11/26 | 12/3 | |
| Work | | | | | | | |
| Work | 7h | 8.3h | 11h | 12h | 12h | 12h | |
| Work | 7h | 8.3h | 5.5h | 12h | 11.5h | 12h | |
| Work | 7.25h | 9.5h | 6.3h | 12h | 11.5h | 12h | |
| Work | 8.5h | 11.3h | 7h | 12h | 11.5h | 12h | |
| Work | 1h | 1h | 1.2h | 1h | 1h | | |

**FIGURE 3.9**
Resource usage hours.

## Process

Rather than look at the detailed tasks, one can also look at the work assigned as a summary by individual. MS Project displays both the total for the project and the hours for particular time periods.

## Example

In Figure 3.9, we see that the total number of hours for all the individuals is approximately equal, although there are variations on a week-by-week basis.

Dr. Knowledge, the advisor, shows only an hour per week, the time for the weekly meeting. Many advisors spend additional time beyond the group meeting on such tasks as finding support or negotiating access to specific tools.

## Resource Costs for Project

| Resource Name | Cost | Baseline Cost |
|---|---|---|
| ⊞ Unassigned | $0.00 | $0.00 |
| ⊞ Sue Swift | $18,905.00 | $18,955.00 |
| ⊞ Joe Bund | $18,905.00 | $18,905.00 |
| ⊞ Wilma Follet | $19,330.00 | $19,405.00 |
| ⊞ Phil Green | $19,055.00 | $19,105.00 |

**FIGURE 3.10**
Resource costs for project.

## Process

The database readily calculates a total of the value of all hours to be spent on a project — extremely useful in developing a business plan in a real situation.

The Baseline Cost column displays the results of a useful feature of MS Project. You may save the project as your baseline. It should be your original, refined plan for the project. You can then modify the project as it develops and compare the revisions to your baseline — often an extremely useful exercise, although sometimes depressing if the optimism level was too high initially.

*Example*

For senior design we see that each of the team members is producing quite a bit — both because of the hourly rate of $100 assigned earlier and because of the number of hours. We see also that at least one task has no resource assigned to it — a milestone lasting zero hours.

### Task Hours over Time

| | ⓘ | Task Name | Work | Details | | T | F | S | S | Oct 1, '01 M | T |
|---|---|---|---|---|---|---|---|---|---|---|---|
| 0 | | ⊟ MS Project Senior D | 773.15 hrs | Work | | 14h | 9.25h | | | 8h | 7.5h | 8h |
| 1 | ▦ | End Spring Term | 0 hrs | Work | | | | | | | | |
| 2 | | ⊟ Start Thinking about P | 8 hrs | Work | | | | | | | | |
| | | Sue Swift | 2 hrs | Work | | | | | | | | |
| | | Joe Bund | 2 hrs | Work | | | | | | | | |
| | | Wilma Follet | 2 hrs | Work | | | | | | | | |
| | | Phil Green | 2 hrs | Work | | | | | | | | |
| 3 | | ⊞ Decide Project Tea | 136 hrs | Work | | | | | | | | |
| 9 | | ⊞ Decide General Pro | 88 hrs | Work | | | | | | | | |
| 14 | | ⊟ Solicit Advisor | 2 hrs | Work | | | | | | | | |
| | | Sue Swift | 2 hrs | Work | | | | | | | | |
| 15 | | | | Work | | | | | | | | |
| 16 | ▦ | Start Fall Term | 0 hrs | Work | | | | | | | | |
| 17 | | | | Work | | | | | | | | |
| 18 | | ⊞ Decide Team Roles | 13.25 hrs | Work | | 0h | 0h | | | 0h | 0h | 0h |
| 23 | | | | Work | | | | | | | | |
| 24 | | ⊞ Develop Project Sc | 28 hrs | Work | | 1h | 2h | | | 2h | | |
| 33 | | | | Work | | | | | | | | |
| 34 | | ⊞ Short Project State | 6.5 hrs | Work | | 5h | 1.25h | | | | | |
| 39 | | | | Work | | | | | | | | |
| 40 | | ⊞ Research | 172 hrs | Work | | 3h | 6h | | | 6h | 7.5h | 8h |

**FIGURE 3.11**
Task hours over time.

*Process*

Yet another way of looking at the core database is to see the details of the tasks showing who is working on them in what time period. Note that the Work column shows the calendar duration rather than the actual hours of effort on the part of the team members — a confusing feature.

*Example*

We see that the group designated Sue Swift to solicit Dr. Knowledge as the team's advisor. It is also clear that for the "Start Thinking about Project" task all the members are sharing the effort.

### Task Cost Summary

*Process*

One last view of the central database allows us to see how the project budget has been spent throughout the project. The view shown in Figure 3.12 summarizes the subprojects, but by clicking on the "+" buttons each subproject could be expanded to view the specific tasks.

| Task Name | Total Cost |
|---|---|
| ☐ **MS Project Senior D** | **$78,335.00** |
| End Spring Term | $0.00 |
| Start Thinking about P | $800.00 |
| ⊞ **Decide Project Teaɪ** | **$13,600.00** |
| ⊞ **Decide General Prɑ** | **$8,800.00** |
| Solicit Advisor | $200.00 |
| Start Fall Term | $0.00 |
| ⊞ **Decide Team Roles** | **$1,325.00** |
| ⊞ **Develop Project Scɪ** | **$2,800.00** |
| ⊞ **Short Project State** | **$650.00** |
| ⊞ **Research** | **$17,200.00** |
| ⊞ **Proposal** | **$20,800.00** |
| ⊞ **Presentation** | **$6,160.00** |
| Proposal Final Due | $0.00 |
| ⊞ **Weekly Advisor Mɛ** | **$6,000.00** |

**FIGURE 3.12**
Task cost summary.

*Example*

We can see that there was probably a mistake made in the hours for "Decide Project Team." This summary would prompt us to review that task in more detail to fix the difficulty. The other aspects of the overall project look in proportion, particularly if one assumes that in developing the proposal, one is doing thinking as well as writing.

### 3.6.2 Illustration of Detailed Tasks

The following screen shots (Figures 3.13 through 3.22) add no further information about the workings of the project management program MS Project. They do, however, present some of the tasks that a design team is probably going to undertake during their startup phase. The later phases would be similar in level of detail but would have different tasks.

### *Deciding the Project Team*

This schedule assumes that a group starts thinking about capstone design in the spring at the end of their junior year, although a number of groups in fact start sooner. Note that they explicitly discuss their individual technical strengths, their quality goals, and their interest in the possible project ideas (Figure 3.13).

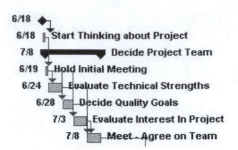

**FIGURE 3.13**
Deciding the project team.

### Deciding the General Project

Once the group is formed, it evaluates the possible projects in more detail, considering not only their interests, but also the possible sponsors and the interests of the potential advisors. While sometimes projects are drawn from the specific desires of a sponsor or advisor, each design project differs in the genesis of the idea, so it is worthwhile to consider the interests of all involved (Figure 3.14).

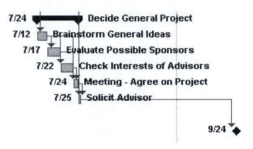

**FIGURE 3.14**
Deciding the general
project.

### Deciding Team Roles

As discussed in considerable detail in the following chapter, it is important to agree on the specific teamwork roles (organizer, recorder, etc.) that are necessary for the success of a project lasting as long as capstone design. These tasks make that discussion and decision process explicit (Figure 3.15).

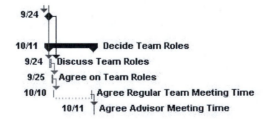

**FIGURE 3.15**
Deciding team roles.

### Short Project Statement

One of the requirements of capstone design is to submit a short, usually paragraph-long, description of the project early in the term. As with most of the other documents for capstone design, a draft and review process will assist doing a professional job (Figure 3.16).

```
9/28        Short Project Statement
9/27      Draft Project Statement
9/27      Review with Team
9/28      Review With Advisor
9/28      Submit to Coordinator
```

**FIGURE 3.16**
Short project statement.

## Develop Project Schedule

In this task, we allocate time to develop the schedule that we have been discussing. Assuming that design project started in September, this task could actually have been begun earlier. It will have the most benefits if the steps of involving the entire team and the advisor in the generation and review are followed. A schedule generated and kept by a single individual is often close to useless except as decoration for the proposal (Figure 3.17).

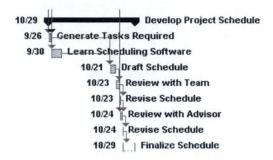

**FIGURE 3.17**
Develop project schedule.

## Perform Research

At last we are into the meat of the design project, with the key tasks of researching. Researching all the factors influencing your potential project is usually essential for real success, but exactly what needs researching will vary from one discipline to another. Further breakdown of these tasks for a particular group is likely to be very helpful. Note, too, the importance of agreeing on the importance of the results. Writing this down becomes a central portion of the formal proposal as well as the presentation of the project (Figure 3.18).

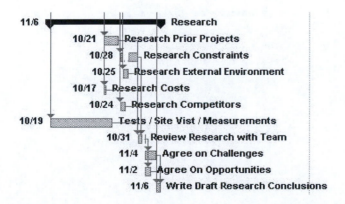

**FIGURE 3.18**
Perform research.

## Write Proposal

The length and requirements for the proposal differ considerably between organizations or departments. In almost all cases, however, there is a need for a main body as well as supporting documentation. Doing a good job on these requires drafts, revisions, and reviews, which take significant time and therefore are specifically indicated by tasks (Figure 3.19).

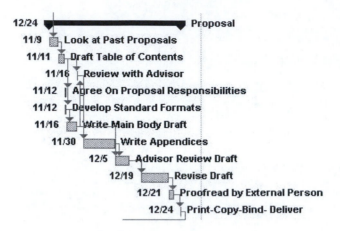

**FIGURE 3.19**
Write proposal.

Although is it not indicated here, there is no reason why some of the work of preparing the written document cannot begin at the start of the first phase: templates, overall table of contents, and the general scope of the appendices can typically be decided very early in the term.

## Develop Presentation

The oral presentation of your project, which occurs at the end of the first phase for most departments, should build readily on the draft of the written proposal. Many of the same materials can be used in both.

As with the proposal, the draft and review processes are extremely important for raising the presentation to a professional level. The complexity of developing a polished presentation is dependent on the teams' familiarity with the presentation software (currently PowerPoint is a popular program), any presentation tools, and the characteristics of the space in which the presentation will be made (Figure 3.20).

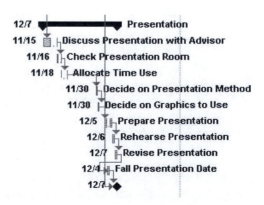

**FIGURE 3.20**
Develop presentation.

This view shows all the tasks that we have seen before, but displayed simultaneously to allow best understanding of their sequence and simultaneity. Note that the solid bars are the subproject indicators that will display if the detail is hidden. This view shows the beginning of the project through the first part of the first phase when the team is formed, the project chosen, the advisor solicited, and the project schedule developed (Figure 3.21).

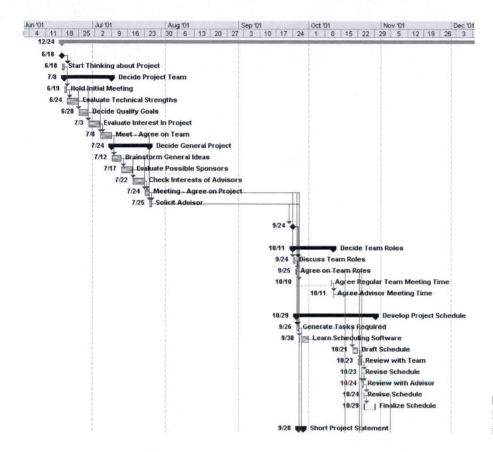

**FIGURE 3.21**
Full Gannt:
initial section.

## Full Gannt: Final Section

This portion of the full project schedule takes us through the end of the first phase. In it the team performs the essential research, drafts the proposal, and prepares the presentation of their project. It is not discussed in detail, but is presented to show all tasks (Figure 3.22).

## 3.7   SUMMARY

This chapter should have convinced you that you schedule all the time, probably without thinking about it. For long, multi-person projects there are many levels of detail at which one may schedule, each with its own level of effort and consequent rewards. The most important thing you can do is list all the tasks that will be necessary to complete the project and then put them in their logical order. Once you have done that you can add how long it takes for each task, who is responsible for it, and how much it will cost.

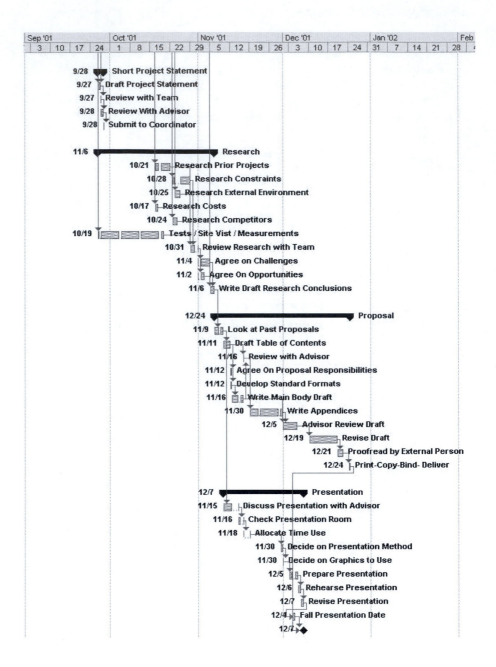

**FIGURE 3.22**
Full Gannt:
final section.

For all but the simplest projects a specialized computer program is beneficial to keep track of the large amount of detail. With each addition of input, you can get increasingly detailed data back from the program that can show a project calendar (Gantt chart), alert you to potential schedule problems, show individual responsibilities, help smooth overload periods, and track costs. Among the greatest benefits of all these reports is the ability to make your plans (and progress) clear to a supervisor or advisor. Often they will have suggestions that may help you through the difficult phases of your project.

# REFERENCES

http://www.eproject.com/products/index.htm (accessed July 19, 2004).

eProject. Although, unfortunately, the free version of this service has recently been discontinued, they will license you an advanced version for a monthly charge.

http://www.imsisoft.com (accessed July 19, 2004).

TurboProject LT. IMSI publishes a number of software programs. They have a version for well under $100.

http://www.microsoft.com/office/project (accessed July 19, 2004).

Microsoft Project. It has links to many help sites. The "project map" that comes with MS Project 2000 has an excellent overview of the entire project management process.

http://www.pmi.org/ (accessed July 19, 2004).

Project Management Institute. The professional organization for Project Managers. There is not too much directly on the site beyond an overview of project management. There is, however, a list of papers on specific topics as well as a few chapters from their handbook of project management.

Project Management Institute. *A Guide to the Project Management Body of Knowledge*: A *PMBOK® Guide*, 2000.

This is the standard handbook of the project management profession, the major reference for their credentialing exam.

# FURTHER READING

Birnberg, H.G., *Project Management for Small Design Firms*. New York, McGraw-Hill, 1992.

Aimed particularly at the construction industry — buildings and civil engineering projects.

Healy, P.L., *Project Management: Getting the Job Done on Time and in Budget*. Melbourne: Butterworth-Heinemann, 1997.

Presents a reasonably compact overview of project management with an emphasis on manufacturing.

★ *Key Points*
*Chapter 4*

- Form the team for necessary skills, not just from friendship.
- Define and fairly share the necessary team roles.
- Understand that there are regular phases of team development: forming, storming, norming, performing.
- Understand conflict and treat it as a potential benefit or at least very carefully.
- Be explicit and careful about communication.
- Develop and continually assess your teamwork skills.

# Chapter 4
# TEAMWORK

*James E. Mitchell*

## 4.1   TEAMWORK AND TEAM ROLES

A major project takes a long time. In college the senior design process, for instance, often takes eight months, while industry projects can range from a week to many years. No matter how strong your liking for your teammates, the project is almost certain to stress your relationship during the dark months of the middle of the project or when that last report is due. Both for the sake of friendships and to maximize the efficiency of the process (less time and higher quality) it is beneficial to address the question of effective teamwork in some detail not only at the beginning, but repeatedly during the project.

### 4.1.1   Talking about Teamwork

In many ways the most important thing you can do is make the subject of teamwork an explicit agenda topic during your meetings. If you are in a business environment you may well have a supervisor who will provide explicit guidance or requirements, whereas in the academic environment you are more likely to shape your teamwork approach yourselves. In either case it is highly beneficial to arrive at a common understanding of the "rules of the game." At initial meetings discuss each of the suggestions of this chapter. If you are in an academic environment you need to elicit each teammate's opinion on each topic and come to a consensus (a vote is less desirable) on the team's attitude toward each. If you are in a business environment you will still need to understand how each of you will work within the requirements set by your management. Discuss the team's decisions about teamwork with your supervisor or advisor. Beyond being able to interpret official requirements, they may have excellent advice based on their own experience that can save you a great deal of time.

Repeat the conversation every month or two, probably in an abbreviated form. You will probably find that the realities do not quite agree with what you projected initially. If everyone is happy and the project is progressing smoothly, accept the change. If not, the conversation can be a way to address difficulties and find a way to resolve them.

As part of that initial discussion you should decide how you as a group wish to handle the various roles that are almost always necessary in a team. It is not necessary that the roles be fixed for the entire length of the project — indeed rotating them will probably help everyone. Some roles, such as "decision maker," have enough prestige that they are overtly or covertly contested. In an academic environment it may well be that the group avoids a decision on these roles while rotating the less desirable but still necessary ones between members. If that is the case you will probably be working in a consensus mode, which requires that all significant decisions (you would do well to decide what *significant* means) be made by mutual agreement of all the members. In the business environment it is more likely that the roles will be fixed once and for all, perhaps by assignment from a superior. It should be pointed out that, particularly in smaller teams, team members must perform multiple roles, and, often, multiple people can perform the same role.

### 4.1.2 Teamwork Roles

There is no universal agreement on the roles necessary for a successful group. The following expanded list makes sense in the context of a capstone design project or early small project in industry. Some experts would reduce this list to only "team leader" and "communicator" or, perhaps, to only "team leader." We have elaborated on the possible roles here because most correspond to tasks (see Chapter 3) that are required in most projects. It often makes sense to have an individual responsible for a given area.

| Role | Meaning | Comment |
|------|---------|---------|
| Analyst | Performs analysis, primarily numeric | Often an individual good with math or a calculation program such as Excel or a particular simulation program appropriate to the discipline necessary for the project. |
| Decision maker or team leader | Makes key decisions | The person who resolves the almost inevitable lack of consensus.<br>In a business setting this individual is usually chosen by management and charged with overall responsibility for the project. The leader has explicit power to assign other team members to specific roles.<br>In academic groups the team may explicitly choose one member for this role. Sometimes the role evolves as the group works together, Or there may be a conscious decision to require consensus for all major decisions.<br>In many academic groups major decisions are made by consensus, while lesser decisions may be delegated either to a "discipline expert" or to a chosen decision maker. |

**TABLE 4.1**
Teamwork Roles

| Role | Meaning | Comment |
|------|---------|---------|
| Facilitator | Helps the group work effectively | Facilitators understand the way groups work. They listen well and are willing to make the effort to understand divergent viewpoints. They attempt to find common ground and to resolve differences through compromise. In many groups this role is not explicitly chosen. Nonetheless, recognizing the importance of the role and rewarding those who fill it will greatly assist the team process. |
| Graphics/models creator | Develops drawings, models, simulations | The specific skills a person filling this role will vary with the project. For those creating a physical object the traditional drawing skills (hand and computer drawings) will be important. For more abstract projects the skill may be more in using software to create process or logic models or simulations. |
| Idea generator | Contributes most significant ideas | In the ideal group the ideas for the project come from the entire group, often through a brainstorming process. In many other groups one or perhaps several members excel at generating solutions to problems.<br>Usually the generation of an idea does not take very long, although the idea person's contribution is critical. Idea-generating members therefore often simultaneously perform other roles as well. |
| Organizer | Develops and maintains process structure | For the group to work well through the length of the project they must stay aware of the necessary tasks and the deadlines, as well as the roles that must be filled. Initially the organizer helps the group formulate the tasks, their interdependence, and the work necessary to complete them. As the project advances, the organizer reviews progress and refocuses effort to meet the deadlines.<br>The organizer is the natural person to take advantage of scheduling/critical path software. |
| Presenter | Presents group information to others | All group members are expected to contribute to any presentation. Nonetheless one or more members may have greater experience or presence. That member can often take a lead role in structuring presentations.<br>In business groups this individual is often the team leader since he or she has responsibility for the project. |
| Recorder | Records the operations of the group | While related to organizing, the role of recorder does not necessarily have to be taken by the same individual. Ideally all meetings should have a written record, emphasizing decisions made and assignments made. This record should be the basis of discussion with the advisor. It also assists in resolution of disagreements later in a project — particularly if a group member does not perform as expected. |

**TABLE 4.1**
Teamwork Roles
(continued)

| Role | Meaning | Comment |
|------|---------|---------|
| Writer | Writes important elements of documents | If the group is lucky, one member will either be skilled at or enjoy writing. This member can take the lead in structuring and writing the critical elements of the proposals and reports as well as any communications outside the group (letters, abstracts, etc.). This individual should not be expected to write everything for the group but should help set the standards and perform editing for consistency and liveliness (yes liveliness is desirable even in a technical report). |
| Idler | Does not do anything much, but a useful label for later discussions | Of course you will not have one, but ... |

**TABLE 4.1**
Teamwork Roles
(continued)

Note that these roles are usually separate from the discipline (for example, electrical or structural engineering) roles that team members perform in a group. For most design projects managers and advisors will expect that each team member perform significant engineering work in addition to whatever teamwork role they fulfill.

## 4.2   ASSIGNING AND ROTATING TEAMWORK ROLES

Each team will evolve a different method of addressing all the project roles — both teamwork and discipline. What is most important is that each team recognizes the necessity of these roles being addressed in some way. Probably the best method is to hold an early meeting in which the primary agenda is the variety of roles and how they shall be covered.

Several useful methods of addressing the teamwork roles may help the discussion:

- Assignment by volunteer: Ask members which role they wish to fill. You may be lucky and find that all the roles are covered without disagreement.
- Rotating assignment: The team agrees that some or all of the roles will be rotated on a fixed time schedule between the members. This usually applies to the less desirable tasks.
- Sharing roles: Desirable roles may be shared, although there is always a danger of lack of clarity about responsibility.

## 4.3   CLARIFY TASK RESPONSIBILITIES AND SCHEDULE

If any issue rivals the importance of being clear about teamwork roles, it is the worth of being clear about the tasks necessary to complete the project and who is responsible for each task. A long design project is so large an undertaking that few if any teams have an initial understanding of all the tasks involved, the relationships between them, and the impact of each on the schedule. The issue is so important and relatively complex that Chapter 3 is devoted to a detailed discussion of how to accomplish worthwhile scheduling.

## 4.4   TEAM DYNAMICS: THE PHASES OF DEVELOPMENT

Most groups go through recognizable stages on the way to becoming an effective team — and some groups never make it. Like much else in the teamwork world there is no universal agreement on those stages, but one definition that makes sense to many was developed initially by Bruce Tuckman (Tuckman, 1965), namely, forming, storming, norming, and performing. This definition argues that before becoming productive, groups must develop their unique method of working together and establishing responsibilities. While it might seem that in a business setting the organization's assigning the team leader would make these phases of development irrelevant, experience shows that they are still usually present, though perhaps not explicitly recognized.

### 4.4.1   Forming

This occurs when the team members are getting to know each other. The most important task is understanding what each team member brings to the process, what they want from it, and how they will interact with each other:

- This stage is often felt as "floundering." Teams often want someone else to tell them what to do.
- It is very helpful to explicitly address the issues raised about teamwork roles in group discussions to make it through this phase.

Often little tangible product results from this phase:

- Discussing the design problem is often the vehicle for testing the teamwork process issues.

### 4.4.2   Storming

Group members are trying out their ideas on each other to test how the group will work and whose ideas will dominate:

- There is often argument and the belief that others "won't listen to me" at this stage.
- Members are more interested in their own ideas than in melding the ideas of all members.
- Recognition that listening respectfully and fully to each member can help greatly.
- Focus on the group process will again help you move through this stage.
- Team members with conciliation skills can be extremely helpful.

The most usual products of this phase are alternatives for the problem statement:

- The team is not yet ready to agree on many details.
- A number of possible tasks or diverse issues may be presented, but not yet resolved.

### 4.4.3   Norming

Groups have learned to trust each other and are moving on to the work of the project:

- Members agree to perform specific tasks (perhaps on a rotating basis).
- The amount of work expected of each team member is agreed on.
- Methods for making decisions are established. "Important" and "unimportant" decisions may use different methods.

The group is now ready to start producing product:

- A common problem statement is adopted and ready for external review.
- A general list of final products can be identified.
- The steps to achieve the product are identified and perhaps formalized in a calendar or Gantt chart.

### 4.4.4 Performing

The group is working effectively as a team. The main focus is on the product:

- Regular meetings are prepared for, attended, and documented.
- Team members perform their roles readily — and perhaps change roles.
- Decisions are readily reached.
- Regular informal and formal coordination between team members allows each to determine if the results of their tasks have affected the work of others.

The group moves through the steps of the design process to a final presentation or report:

- The results of tasks are documented and shared.
- Drawings, reports, and prototypes are produced efficiently and effectively.

## 4.5 CONFLICT RESOLUTION

Disputes or disagreements are inevitable when you have more than one person working on a shared task. The best groups use disagreements as a source of inspiration for novel solutions. The worst disintegrate into name-calling and disaster.

In many instances the conflict first becomes apparent when the project appears likely to fall behind schedule or to be delivered incomplete or late (if there is no schedule). A number of useful techniques for addressing those issues are contained below in Section 4.6.

The suggestions below should help you make the best of disputes that do not appear to be based on schedule problems.

### 4.5.1 Act Calmly

- In disputes tempers and voices naturally rise. Consciously talk calmly.
- Listen to what the other person says.
- Assume that others are well-intentioned until proven otherwise.
- Sometimes letting an issue settle overnight will be enough.

### 4.5.2 Clarify the Dispute

If it is a matter of intellectual disagreement, then rational approaches using examination of the merits of the different opinions is likely to be helpful. Think of using:

- Explicit statement of the facts that are under discussion — there are often quite erroneous perceptions.
- Comparison of criteria.
- Consideration of the consequences of each approach.
- Examination of similar situations and what was done there.

### 4.5.3 Look for Compromise

If it is a matter of personality differences or a disagreement about the amount of work performed, resolving the dispute is often harder. It is certainly worth considering the same techniques as for intellectual disagreements. It is often helpful in addition to:

- Use a third party mediator — another team member
- Have a cooling-off period
- Discuss the feelings involved — usually with an agreement to let each person speak his or her piece while the other listens in silence

Often there is a compromise between two opinions (particularly the intellectual ones) that will be better than either of the two alone. For personality disputes the compromise may leave each person feeling the resolution was fair even if they did not get everything they wanted.

### 4.5.4 Use an External Party

If you are unable to resolve an issue within the team, it may make sense to appeal to a respected third party. That may be the group's supervisor, or possibly an individual not in the chain of command. In business it is likely to be the immediate supervisor, but not necessarily. For individuals not in the chain of command it is critical that they be regarded as fair and knowledgeable.

## 4.6 WHAT TO DO WHEN YOU FALL BEHIND

While some projects proceed without delay, confusion, or last minute efforts, they are few. Delay usually occurs for one of four reasons: tasks or task duration was not estimated well, customers or team members changed their minds, team members or customers did not communicate well, and team members procrastinated. We will examine each of these causes separately and suggest some useful tactics for both avoiding and, as is often sadly necessary, dealing with the crisis.

As a general rule, the best way to deal with falling behind is to avoid it. You should try to anticipate potential problems and make allowances for these by leaving extra room in your schedule or your budget. Techniques for dealing with unknowns are discussed in Chapters 1, 3, and 5.

One key consideration made for any project that falls behind schedule is whether it is possible to delay completion. For student projects the deadlines are often fixed by the end of courses and cannot be changed. In that case the usual student answer is to work more intensely at the end of the project — the infamous all-nighters. In industry many projects are promised on specific deadlines, with major financial or even life-safety consequences if the schedule is delayed. If that is the case, then overtime, however unpopular or expensive, may be necessary. In the long run it is much better to advise management as soon as a problem is definite — most managers hate last-minute bad news.

However, there are many projects in which delays can be tolerated, though almost never appreciated. Again, it is much better to address the delay and receive approvals as early as necessary.

In all of what follows we are assuming that a delay in a final deadline is not acceptable and that the team must find ways to meet the deadline, producing a promised set of deliverables.

### 4.6.1 Unanticipated or Underestimated Tasks

Very common problems for first-time project participants or for those working with new technology are not anticipating necessary tasks or underbudgeting time and other resources for tasks. As always, the best way to address the problem is to anticipate it — in this case that means building extra time into the overall project schedule. Some useful techniques for addressing the problem include:

- Looking for tasks that can be eliminated or completed after the deadline
- Looking for ways of making acceptable compromises on the extent of promised deliverables
- Working longer hours, with the inevitable consequences of lost sleep, financial costs, and degradation of efficiency
- Prioritizing the deliverables and producing only those that are possible within the available time, accepting whatever consequences occur for incomplete performance as better than the penalty for delay

### 4.6.2 Changed Minds

In a design process everyone, the client or user included, evolves over time both because of the increased knowledge that develops as a project design emerges and because of changes in the project's environment (for example, a change in company objectives or profitability). What is important when the project starts may not be appropriate halfway through the design process.

As always, planning is the best way to address this problem. At the beginning of the project define and get written approval from your client, supervisor, or advisor for all the critical design assumptions. These assumptions include the deliverables, the finished product of your design effort. They include the conditions in which it is expected to operate, the production and operation environments, the resources you will have available during design, the time to receive intermediate approval — all the things that can affect the schedule if people's minds change during the design process. Different disciplines have different names and times for documenting these ground rules, but in almost every case the effort to be very explicit before you start design will pay off handsomely.

It is also extremely important to agree within your group on the dates at which decisions about the design are to be fixed and unchangeable. Almost every designer wants to find a better solution up to the last minute. If you allow such changes you are highly likely to fall behind.

If, despite your best efforts, someone changes his or her mind, then the following techniques may be helpful, both before and at the crisis point:

- Document all decisions when they are made. By doing this you make later discussions of consequences of changes not dependent on highly fallible memories.
- Discuss the consequences of changes with those requesting them. If you have documented decisions, you can demonstrate that a change violates a prior agreement. If the individual insists on the change anyway there can then be decisions on appropriate consequences such as extensions of the time and increase of the fee if the client has changed their mind. If a team member has changed his or her mind the team can overrule the team member if the change is minor or, if it is a critical change that must be made, can have a basis for assigning extra work.

### 4.6.3   Ineffective Communication

This problem is closely related to the issue of individuals changing their minds, discussed above. Communicating effectively both orally and in writing is extraordinarily important in keeping a project moving smoothly. Far too often many ideas are discussed in a vague way, with individuals taking away from the conversation what they want to hear. Only by being explicit about the agreement and documenting it in clear writing is it likely to be understood at the time and later when the decision needs to be reviewed.

Steps you can take if a project is delayed include:

- Review the written record to see if there are prior decisions that can provide a solution to the current problem.
- Review the written record to see if it provides a process for resolving the problem.
- Prepare a clear, written summary of the problem for discussion with the relevant individuals. Often the process of preparing such a summary helps point out the best solution.
- Using that summary as a basis, discuss the problem face-to-face with those who can provide a solution. Perhaps it is only your team members, but often it is those beyond the team. Face-to-face discussion is usually vital for significant problems; there is too much that the interplay of all the communication modes provides (body language as well as oral and written) that cannot happen via an e-mail or telephone call.

### 4.6.4   Procrastination

Procrastination is inevitable and occasionally beneficial (it can give time to assimilate and understand complex relationships). As with the prior sources of problems, the best solution is to plan ahead via a detailed schedule that is agreed to by all the team members. By doing so you can know not only when each individual is supposed to have provided final product, but also when they are supposed to begin and when they should need input from others. Failure to meet these intermediate deadlines can be a critical signal of procrastination that can be addressed before it becomes serious.

If individual procrastination leads to a problem, it often appears initially in one of the three problem categories discussed above. The solutions for them are still valid, probably with additional attention to individual responsibility and taking measures to avoid future recurrence. The measures that are appropriate especially to procrastination difficulties include:

- Discuss the consequences of nonperformance with the individual. It may be that the individual does not realize what effect his or her lack of performance at a given point is having on the team's overall work.
- Put the individual's nonperformance on the agenda for a group meeting. Just the threat of public exposure may be enough to resolve the problem.
- Document the problem. It is possible that the individual does not realize that the problem is real. By developing a record you can make the problem unavoidably clear to the person responsible and, if necessary, to a higher authority.
- Change a person's role assignment. Since some roles are almost always more desirable than others, it may be possible to switch a person from one role to another. Often that role would be one that can be more closely supervised.
- Appeal to higher authority. A supervisor or advisor may use the threat of unpleasant consequences to produce performance. This is usually the last step one wishes to take since it could easily sour the working relationship in a team.

## 4.7  COMMUNICATIONS

### 4.7.1  Hold a Regular Meeting: With Food

It should be obvious that meeting as a full group on a regular basis is essential for a successful project. The groups that have done well in capstone design projects in the past have treated those meetings as critical. For most groups the characteristics of those meetings have been:

- Held at least once a week in addition to the meeting with the advisor or supervisor. Sometimes these meetings should be held twice a week depending on the urgency and nearness to a deadline.
- Held at a regularly scheduled time so that each member can predict and schedule the meeting.
- Includes all team members at each meeting. Someone missing more than one or two meetings is a good indication of trouble developing in the team.
- Includes food of some sort to establish a relaxed atmosphere.

### 4.7.2  Run a Successful Meeting

Too many (most?) meetings are wasteful and boring. They are, nonetheless, essential to be sure that the team works together. Agreeing how to run your meetings to be short and effective will save both time and tempers. You are almost certain to have informal meetings as well, but for students you are wiser to make these social meetings with side conversation about the project. Save important decisions for formal meetings.

There are four essentials for running a good meeting — apparently simple but taking considerable effort to achieve in practice:

- **Have an agenda:** This should be prepared in advance, probably by the organizer after review of the prior meeting minutes and polling of those coming to the meeting for new issues.
  - Organize the agenda so that the most important issues are resolved first — that way anything that is not complete can be allowed to wait till the next meeting.
  - Eliminate anything that can be resolved by e-mail or one-on-one communication.
- **Listen to everyone:** It is almost inevitable that one or two members of the group will be more vocal than others. If you are to keep the silent ones involved and committed it is essential that you hear everyone's opinions. The facilitator is particularly important to monitoring the conversation so that each person is heard. It may be important to agree in advance that you will have an informal time limit for comments, to give time for each person.
- **Stick to the agenda:** This is usually the role of the organizer — politely but firmly reminding the group of their prior agreements and the passage of time.
  - Assign a time to discuss items is often very helpful.
  - Agree in advance what to do if discussion of an item is not complete.
  - Finish on time — your friends will remain your friends.
- **Keep good records:** so you do not spend time repeating discussions (see the section that follows on methods of keeping good records).

### 4.7.3 Keep Records of Team Meetings and Agreements

Human memory is fallible. Your supervisor or advisor may require you to keep minutes of team meetings. Even if they do not, you would be wise to learn from centuries of experience that keeping a record of at least the decisions and task assignments made at each meeting is highly beneficial. First, it is a reminder of who is responsible for what aspect of the project that can be reviewed at any time. Second, if there is a dispute of some sort, a written record exists to trace the history — and to show the advisor or supervisor if that sad necessity arises. Third, the process of writing down a decision often illuminates its relationship to other tasks in process.

There are a variety of ways of circulating and storing the records. Probably the best is to take advantage of a Web service like eProject or a shared network with a folder holding all meeting minutes. The recorder can assign a task (or send an e-mail) whenever a new meeting minute is put in the folder.

### 4.7.4 Assess the Process

Periodically discussing not only the technical issues of design, but also the overall process of the team and the manner in which members are meeting their commitments can avoid later strife. A useful tool would be a copy of the list of teamwork roles defined above. If members independently fill out their assessments of what roles they have filled and what roles they perceive the other team members as filling, it can be the basis of a very productive discussion.

### 4.7.5 Communicate with the Outside World

When communicating with people outside the team (clients, sponsors, code officials, manufacturers, etc.), remember that they may be important to you in the future. Always conduct such communications in a professional manner, as though you were representing your firm as well as yourself — which you are. Particular issues to remember include:

- Use formal titles and correct addresses in anything written.
- Be sure to identify yourselves clearly. It is particularly important that anyone you contact understand your responsibilities and role — as a student or professional. For some types of projects there could be very serious legal consequence for misrepresentation.
- For student projects never promise a sponsor the results of a project for a fee. Instead make it clear that they are sponsoring you in a scholarship manner and that you may or will provide them with a copy of your results as a courtesy.

### *Electronic Communication*

Electronic communication assumes an ever-increasing role in any project. Indeed some participants in professional projects complete highly successful projects without ever meeting (although the continued growth of conference rooms at airports testifies to the important role of face-to-face meetings). There are some issues worth considering about each method, including both politeness and efficiency.

### *E-mail*

Almost every student and professional is now proficient with e-mail. Nonetheless some general suggestions may be helpful:

- Use it between meetings — it is great to keep a project moving.
- Be sure to copy all concerned.
- Save any e-mails containing important information or decisions as separate files — e-mail is notorious for disappearing at the wrong time.
- Consider using a dedicated project management service such as eProject, which incorporates ready e-mailing to all participants at their regular e-mail address.
- Watch out for "flames" — rapidly escalating exchanges of insults, which often grow out of a simple misunderstanding that would be resolved in ten seconds in a face-to-face meeting.

### *Telephone*

Remember its ephemeral nature. If you make a decision or receive some information during a telephone call be sure to write it down and send an e-mail or paper copy to everyone concerned if an important decision or piece of information has been exchanged.

### *Shared Folders*

Deciding on a central location for all electronic documents can greatly increase team efficiency. Many services now provide that ability including IDrive and eProject. When using this kind of sharing it is wise to consider the following issues:

- Upload and download time for files: Some files may just be too big for centrally storing and ready use at home.
- Naming conventions: Agree on how files will be named so that everyone can rapidly understand what is in a file. A date as part of a file name can be particularly helpful.
- Folder organization: As with naming conventions, agreeing on a hierarchy of folders can make retrieval of the correct information much easier.
- Checkout conventions: Agree on the methods for ensuring that more than one person does not alter the same file simultaneously.

### *Web Pages*

A project Web page can be extremely beneficial when communicating with multiple groups or as a final product documenting your work. Remember that without security, Web pages, shared folders, and Web services may be visible everywhere on the Web. Make sure to treat your electronic communication with the same care that you give to any communication with the outside world.

## 4.8   SUMMARY

Teamwork is obviously one of the most important ingredients of a successful project and one for which there is often little explicit training. It can help greatly to understand that there is a four-step process to forming truly effective teams. When each member understands that the first steps are primarily about building relationships and establishing roles within the team, then a lack of initial outward progress can be less worrying so long as the group understands that they must achieve the performing stage to actually produce a product.

While there are no rules as regular as those in physics for how to function truly efficiently in a team, there are useful techniques for ensuring that communication happens and is documented appropriately. There are also useful techniques for addressing the problems that arise almost inevitably during a long project. In particular there are ways of addressing conflicts between team members and also of taking beneficial steps when a team falls behind.

## REFERENCES

Tuckman, B.W. (1965). Developmental sequence in small groups. *Psychological Bulletin*, 63, 384–399.

> Elaborates on this model of teamwork development as well as several other models. This model can also be found at a number of Web sites including http://all.success-center.ohio-state.edu/all-tour/director.asp (accessed July 19, 2004).

## FURTHER READING

Most books on project management address the teamwork issues explicitly. See the scheduling chapter (Chapter 3) for suggestions.

http://files.irt.drexel.edu/courseweb/Mitchell_Courses/Teamwork/. *Teamwork: A Practical Guide for Students* (accessed July 19, 2004).

> This site was developed for individuals working on their first long project. It is aimed at college students but should be applicable to graduates early in their careers.

★ *Key Points*
*Chapter 5*

- Every project must be evaluated for its economic feasibility.
- You can develop several types of budgets for your project:
  - A budget for the actual cost to you (or your sponsor) of creating your initial design or prototype, excluding salaries, benefits, and any free services or equipment available to you from your university or department.
  - A budget reflecting the entire cost that a company would incur to support the engineering development efforts required to design your product, produce a prototype, test the prototype, iterate the design, and complete the required documentation.
  - A budget that would be part of a business plan, which would cover engineering development costs, marketing and sales costs, and manufacturing costs and include the projected time to recover the development costs.
- Your prototype budget should be carefully developed and then reviewed throughout the duration of the project.
- Contingency plans for budgets and schedules are vital to the success of your project.

# Chapter 5

# ARE WE IN BUSINESS YET?

*Stewart D. Personick*

## 5.1 INTRODUCTION

An essential part of any business activity is preparing a budget, or a set of complementary budgets. Bringing a new product or service to the marketplace requires an increasing investment of money until you reach the point where the monthly income from sales exceeds the monthly expenses. Budgets are tools for estimating such things as what those required investments are going to be, when incremental monthly sales revenue will exceed incremental monthly investment, when the cumulative income from sales will exceed the cumulative investment (if ever) and for evaluating progress vs. plan during the execution of a project.

Typically, for a senior design course, you will be required to produce a series of budgets and to perform an economic viability study for your proposed product or service. For an industrial project you can usually assume that your customer or manager has determined that the project has benefits and you will only be required to produce a budget for a concept-demonstration prototype and final product. Thinking about a budget early on allows you to determine whether your design project is economically viable. If you propose to design and build a concept-demonstration prototype of a new type of widget, you must be reasonably sure that you will be able to obtain access to the resources you need to develop your prototype and that the required technology (for example, components and software) is available to you at a price you are prepared to pay.

The following is an example of a problem that might arise, and which could be anticipated in the budgeting process. Design teams sometimes become excited about a new technological development they have read or heard about (for example, a new kind of high-temperature, super-conducting sensor or a new type of optical switching component), and they come up with a creative way to utilize the new technology in a concept-demonstration prototype. However, if a technology is both new and exciting it may also be very difficult to obtain sample components that embody that technology. Other people who are trying to develop and sell new products will likely be excited by this new technology as well, so there may be a large backlog of orders for components as initial demand exceeds the supply. In addition, people who are willing and able to place large orders will be given priority over students looking for a few samples. As a result, you may be required to pay a premium, often much more than the list price, to obtain the needed sample components in time to complete your project, or you may not be able to obtain any samples at all. Preparing a budget, and asking your advisor or manager to provide feedback on that budget, may help you identify potential cost-related problems before they arise.

The process of developing a budget and comparing it to others may also help you and your team to identify all of the required resources, other than components, that you will need to successfully complete your project. If you will need a specialized CAD tool for your engineering design and simulations, you must determine, ahead of time, whether it will be available to you (for free) through your department or from other sources and, if not, how much you will have to pay to obtain access to it. Similarly, you may need large blocks of uninterrupted computer laboratory time or access to specialized test equipment in the development of your prototype. You must check with the relevant laboratory managers to determine whether those resources will be available to your team, and you may need to schedule, in advance, your team's access to those resources.

Finally, it is important to gain some experience in the methods that are used to determine whether your product or service could ever be economically successful in the marketplace. If it will cost thousands of person-hours to develop the final product, and if the product would be expensive to manufacture and support, will you ever be able to recoup those development and manufacturing costs if you proceed to sell the product in the real world to real customers who have the alternative of not buying your product? Developing a budget lets you see, in advance, the approximate costs of development and manufacture of a proposed product and allows you to make engineering design and business-related trade-offs that may make the resulting product economically viable.

Business is simple ... you agree to do something, and then you do it.

**President of a successful $5 billion/year company, circa 1979**

Some people's problems seem to follow them around.

**Senior executive of a large engineering firm, circa 1975**

Since we have used such words as *estimate* and *approximate* above, it is appropriate to mention that in the real world of business, successful engineers deliver on their commitments by consistently completing design and development projects within the budgets they have produced or agreed to meet for their projects. In the real world of business, everyone has problems to deal with. In the real world of business, nobody cares about the reasons you can provide for why your project exceeded its budget. You are expected to anticipate problems and to take steps to offset the effects of problems by taking advantage of a combination of resourcefulness and good judgment, at the outset, in creating your budget.

## 5.2 TYPES OF BUDGETS

You will generally be required to produce several complementary budgets.

The first budget will be the actual cost to your customer, your team members, your department, or any academic and industry sponsors of producing your concept-demonstration prototype. Your concept-demonstration prototype is what you will demonstrate either at the end of the prototyping phase of your project or at the completion of your design project (if all you are required to produce is a prototype).

This first budget will include an itemization of the concept-demonstration prototype's design, construction, testing, documentation, and demonstration costs that you expect to incur. Usually this budget will include an estimate of the number of hours each of your team members will have to expend on each of the various tasks (that you have defined) that make up the total project. For a senior design course you will value the cost of those hours at zero (since it is assumed you will work for free on your senior design project during your senior year). While you will not have to pay for your own time, your time is still a limited resource, and you need to estimate how much of that limited resource will be consumed by each task you need to accomplish. Also, because the cost of people's time is a major component of any real-world project's overall cost, it is good to get some practice in budgeting the hours of time required to perform various types of tasks. However, for an industrial project you must produce an accurate estimate of the time and salaries required for the duration of the prototyping phase.

If you have to pay someone who is not a member of your team to do circuit board production, machining of custom components, testing, or to provide other services, these costs must be included in your budget. Even if you expect these services to be provided to you for free, it is a good idea to list them as cost items, with an associated cost of zero. That way, your budget will not appear to be missing important items that would normally contribute to the cost of your prototype. You must include all significant, required items of software and hardware and computing or laboratory facilities that you expect to be available from your department, company, customer, college, university, or outside sponsors that would normally contribute to the cost of your concept-demonstration prototype, even if these will be available to you at no cost, as part of your laboratory fees, an existing grant, or a customer agreement. As with donated services, those items that you expect to obtain at no direct cost to your team should still be listed as line items on your budget. You should indicate (perhaps with a footnote) that these will be available to your team at no cost, and you should list the associated cost for each of these items as zero.

For a senior design project, you must also indicate how you are going to pay for all of the costs your team will incur. If the total of those costs is low, then you can propose to split these between the team members as your out-of-pocket contribution to your project or you can determine whether your department has a budget for senior design expenses. However, it might cost $75,000 to design, build, and race a solar powered electric car. When the cost of building the concept-demonstration prototype is beyond your team's ability to fund out-of-pocket you will need funding from your advisor, the university, or an industry sponsor to complete your project. Without an identified, credible source of the required funds, your project proposal is incomplete.

For an industrial project your projected expenses must be vetted by your managers, and often by your second-line manager (their bosses). Therefore it is necessary to carefully determine appropriate costs for each item and determine whether any of the components or services can be obtained at reduced or no cost.

The second budget or part of a budget produced in a design project will typically reflect the cost of developing a manufacturable prototype of your product or service in an industry setting. It will include such things as the engineering personnel–related costs missing in the concept-demonstration prototype budget and it will include overhead costs to pay for the use of facilities and support personnel. Typically this second budget would also include costs associated with creating engineering documentation needed for manufacturing your product (or delivering your service) and for supporting customers who purchase the product or service. This product development budget will be discussed in Section 5.3.

A third budget that will be included in many senior design projects is associated with an overall business plan that takes into account all revenues (income from the sale of the product or service) and expenses from initial prototype development through the recovery of cumulative expenses. This is typically not required at the engineering level in industrial projects, since teams are typically assigned projects by departments or people who have already performed this analysis.

In the following section, project expenses are discussed and examples of budgets are given. These are representative budgets and may not reflect all of the expenses you will encounter during your design project. You should work with your manager or advisor to develop a realistic set of budgets and develop contingency plans for what to do in case costs grow to be more than anticipated.

## 5.3   PROJECT EXPENSES

If you have started to think about what your project will cost, you have already realized that there are many categories of expenses you could incur. These range from the obvious costs of purchasing components for your prototype, or software to assist with design and analysis, to the often hidden costs of overnight mail to get your components on time or photocopying costs for the multiple copies of the reports you are required to submit.

In general, product development expenses (not limited to just the concept-demonstration phase) fall into several functional categories:

- **Engineering development costs:** the costs you incur to design the product or service, produce a prototype, test the product or service, iterate the design based on the test results, and document the design in sufficient detail to enable the product or service to be produced, delivered, installed, and supported
- **Marketing and sales costs:** the costs you incur in understanding what features and functions the product must have to meet customers' needs, understanding what competitive products or services are on the market, creating a sales plan, and implementing the sales plan
- **Production costs:** the costs you incur in reproducing the product or service for delivery to your customers
- **Support costs:** the costs you incur in helping your customers solve problems with your product or service and providing warranty services to your customers
- **General and administrative costs:** the costs you incur for such things as legal and accounting fees and general management of your company

Within each of these functional categories, there are:

- **Labor costs:** the hourly or monthly salaries that you must pay your employees
- **Materials costs:** the cost of purchasing materials
- **Services costs:** the cost of purchasing services (for example, using another company to design or manufacture something you need or advertising expenses)
- **Travel and living expenses:** expenses associated with business travel
- **Fringe benefit costs:** the contributions to employee pensions, health insurance, vacations, and sick days that you pay for
- **Overhead costs:** the charges assessed on all projects to recover miscellaneous expenses incurred by the company or firm: the rent on the building your company rents space in; the monthly bills for electricity, water, heating, garbage disposal, snow removal, and security services; telephone and data communication services; cleaning services; the cost of secretarial and other administrative staff; expenses incurred for research activities; expenses associated with the salaries of employees for hours when they are not assigned to useful projects; employee training; etc.

We focus on the first functional category, engineering development costs, since this is the category that will be the focus of your two main senior design project budgets. However, to complete a full economic viability analysis for a product or service you must consider all of the above functional categories of expenses and balance them against your potential sales revenues.

### 5.3.1 Engineering Development Costs

As part of any new product or service development, the engineering department of a firm will be expected to produce an engineering development budget. An example of an engineering development budget is given below. In case you did not notice, engineering development projects are expensive.

Estimates of hours required for various tasks are usually based on experience with similar projects that have been conducted in the past. A common mistake is to be overly optimistic (perhaps unrealistic) about the time it takes to design, test, redesign, retest, and document a real product that will be produced and sold to customers. This is notoriously true in the case of software. One must be particularly careful to distinguish between the design of a real product and the design of a concept-demonstration prototype. A concept-demonstration prototype (such as you will create in your senior design project or in the initial phase of your industrial project) may look like a real product and may perform some of the functions of the real product it is intended to demonstrate the concept of. However, it is not unusual for the time and cost associated with the design of a real product to be 10 times as great as for a concept-demonstration prototype. As an example, a real product that utilizes digital circuitry must be certified to be compliant for Federal Communication Commission (FCC) "part 15" radio frequency emission limits. This compliance can be very expensive to test and difficult to meet as a design requirement. A real product may have to work reliably over a wide ambient temperature range. Concept-demonstration prototypes are rarely expected to take those kinds of requirements into account.

In the following example the costs of all components for the prototype are consolidated into one line in the budget. This is because the total of all anticipated component costs, in this particular example, is relatively low compared to the

| Category | Expense | Cost per Unit | Total Units | Total Cost |
|---|---|---|---|---|
| Initial design | Hardware engineer<br>Software engineers (2)<br>Project manager | $35/hour<br>$35/hour<br>$50/hour | 400<br>800<br>400 | $14,000<br>$28,000<br>$20,000<br>$62,000 labor subtotal |
| Prototype construction | Hardware engineer<br>Software engineers (2)<br>Project manager<br>Technician | $35/hour<br>$35/hour<br>$50/hour<br>$25/hour | 200<br>400<br>200<br>200 | $ 7,000<br>$14,000<br>$10,000<br>$ 5,000<br>$36,000 labor subtotal |
| | Components for three prototypes (includes: subcontracted, custom-made assemblies) | $10,000/each | 3 | $30,000<br>$30,000 materials subtotal |
| Prototype testing | Hardware engineer<br>Software engineers (2)<br>Project manager<br>Technician | $35/hour<br>$35/hour<br>$50/hour<br>$25/hour | 40<br>80<br>40<br>80 | $ 1,400<br>$ 2,800<br>$ 2,000<br>$ 2,000<br>$ 8,200 labor subtotal |
| Design iteration | Hardware engineer<br>Software engineers (2)<br>Project manager | $35/hour<br>$35/hour<br>$50/hour | 40<br>80<br>40 | $ 1,400<br>$ 2,800<br>$ 2,000<br>$ 6,200 labor subtotal |
| Prototype modification | Hardware engineer<br>Software engineers (2)<br>Project manager<br>Technician | $35/hour<br>$35/hour<br>$50/hour<br>$25/hour | 40<br>80<br>40<br>80 | $ 1,400<br>$ 2,800<br>$ 2,000<br>$ 2,000<br>$ 8,200 labor subtotal |
| | Components | $15,000/each | 1 | $15,000<br>$15,000 materials subtotal |
| Modified prototype testing | Hardware engineer<br>Software engineers (2)<br>Project manager<br>Technician | $35/hour<br>$35/hour<br>$50/hour<br>$25/hour | 40<br>80<br>40<br>80 | $ 1,400<br>$ 2,800<br>$ 2,000<br>$ 2,000<br>$ 8,200 labor subtotal |
| Documentation | Hardware engineer<br>Software engineers (2)<br>Project manager | $35/hour<br>$35/hour<br>$50/hour | 120<br>240<br>120 | $ 4,200<br>$ 8,400<br>$ 6,000<br>$18,600 labor subtotal |
| | | | | $147,400 total direct labor<br>$ 45,000 total materials<br>$ 51,590 Xfringe benefits on direct labor (@ 35%)<br><br>$243,990 total direct and fringe benefits<br>$304,987 overhead (@ 125%)<br><br>$548,977 grand total engineering development cost |

Example of an
Engineering
Development Budget
for *The Voice-Activated
Appliance Controller*

cost of labor. If we were, instead, proposing to build a prototype with a larger portion of its total cost associated with components, then all of the expensive components should be listed individually. Often, one will need a written justification for the selection of each of these expensive components; that is, higher-level management will want to know the engineering reason why expensive components are required (vs. lower-cost alternatives). In industrial product development, engineering teams are expected to produce detailed specifications for any required components that are not standard components that are available "off-the-shelf" from multiple component suppliers. These detailed specifications are used, by the purchasing department, prior to the production (manufacture) of the product to produce requests for quotes from potential component suppliers. Then they are used to select qualified suppliers and to produce purchase orders for the components.

These specifications are also used to verify that the purchased components delivered by the suppliers are acceptable. If a component is not available off-the-shelf from multiple component suppliers, then the engineering team is expected to work with the company's purchasing organization to identify two or more qualified sources of those components. For example, if the product requires components that can work properly at temperatures from –30 to 90°C, then the engineering team may be required to work with the purchasing organization to identify component suppliers who can meet those requirements, and the engineering team may be required to develop special testing equipment to be used by the product production organization to test purchased components for compliance to specifications. All of this translates into engineering tasks in a real product development that need to be budgeted.

In the engineering development budget example no costs were explicitly shown for purchasing or renting equipment. We are, by implication, assuming that this project will only utilize equipment that is already available on-site to the project development team members. We are also, by implication, assuming that this equipment is purchased and maintained by the company, that its cost is included as a component of the company's corporate overhead rate, and that it is shared between projects on an as-available basis. Therefore, we may need to schedule our use of the equipment in advance to ensure that it will be available to us when we need it.

If you require special equipment that is not already available on-site, the cost of this equipment must be added to the budget, and written justifications supplied; that is, companies are reluctant to invest in equipment that may only be used for one project. A justification is needed to explain why the project team really needs this equipment (vs. existing equipment that is already available to the team) and why this equipment should be purchased rather than rented. In some companies, equipment is purchased and maintained on a corporate basis, but that equipment is rented to projects that utilize the equipment. In that way, projects that use the equipment incur the equipment costs, and the corporate overhead rate is lower because these equipment costs are not recovered using overhead.

Overhead rates vary widely between companies and are dependent upon whether certain costs (for example, secretarial salaries, training time, administrative meeting time) are charged directly to a project or are considered part of overhead. Typical overhead rates are between 100% and 175%, with higher numbers considered indicative of a company with poor cost controls. University overhead rates are generally much lower than the overhead rates in industry, for a variety of reasons. For example, universities typically use cost accounting systems that charge all paid hours, expended by faculty and students, directly to projects, rather than spreading the cost of unproductive activities over all projects, as overhead. In industry, time spent by an employee at an all-day department staff meeting would typically be charged to overhead, rather than being charged directly to any of the projects that engineer is working on.

### 5.3.2 Concept-Demonstration Prototype Development Costs

For a senior design project, even if you have proposed to develop *The Voice-Activated Appliance Controller,* as in the previous example, you obviously are not going to spend over half of a million dollars on the project. You are creating a concept-demonstration prototype as part of a senior design project and therefore will not have the personnel, fringe, or overhead costs associated with an industry project. (Of course there are costs in maintaining the laboratories you work in and the university medical service, but these are covered by your tuition, so we will not count them). However, you will still encounter expenses, and it is necessary to ensure that you are aware of them and know how you will cover them before you spend much time on a project that will be too expensive. If you are participating in an industry project, this is an example of a true budget and illustrates why all of the design steps of Chapter 1 are necessary in order to restrain project costs.

Let us consider an example of a budget associated with a project in the Appendix. The associated sample budget is given below. The budget for *The Talking Book* shows the expenses (actually incurred) by the team in building their senior design concept-demonstration prototype.

| Actual Expenses Incurred to Date | |
|---|---|
| Microcontroller evaluation programming unit | $72 |
| Three reprogrammable microcontrollers | $54 |
| Miscellaneous electronic components | $150 |
| RFID transceiver and tags | $1,537 |
| Telephone | $150 |
| Office supplies, printouts, shipping, misc. | $150 |
| Total | $2,113 |

There are several things to note about this budget. First, it is an overview of a full budget; more detailed information was placed in an appendix. Second, it is retrospective, that is, it was revised as the costs of the project were updated to take into account new or updated information; however, the team's final project costs were less than the initial estimate, so they estimated costs well.

Consider *The Talking Book* as illustrated by the draft of the final project report in the Appendix. Since, at the time of the project, radio-frequency identification (RFID) technology was new, it was expensive. Therefore, the cost of the RFID transceiver dominates the budget. However, even miscellaneous electrical and office supplies were significant out-of-pocket expenses for the team, and costs of calling suppliers and equipment support personnel total 7% of the budget! This shows that it is easy to underestimate what one will have to pay for a project.

*The Talking Book* project was complete at the time the sample final budget was reported by the team. It does not reflect many of the true costs associated with the project, just the team's out-of-pocket expenses for hardware and supplies that were not provided by the university.

Let us now examine a budget, which could represent projected expenses at the beginning of this project. Note that this is a modified version of the budget that was produced by *The Talking Book* project team. Observe that the cost projections are slightly different than the final (actual) project costs shown above. Hopefully, as this team did, you will be able to make solid projections so that you do not have any surprises. As with scheduling, the ability to make reasonable cost projections comes with experience.

| Expense Category | Costs |
|---|---|
| Electrical components | |
| RFID transceiver and 10 tags | $ 2,200 |
| Custom manufactured and tuned antenna | $ 200 |
| Four reprogrammable microcontrollers | $ 80 |
| Batteries and other miscellaneous components | $ 150 |
| Equipment | |
| Laptop | $ 2,000[1] |
| Microcontroller evaluation programming unit | $ 100 |
| Power supplies (2) | $ 150[2] |
| Signal generator | $ 6,000[2] |
| Logic analyzer | $10,000[2] |
| EEPROM eraser | $ 60[2] |
| Software | |
| PSpice for circuit design and simulation | $ 4,095[3] |
| Supplies | |
| Phone charges | $ 50 |
| Photocopying and binding of reports | $ 40 |
| Shipping | $ 50 |
| | $25,175 Total project cost |
| | $16,210 Cost of equipment and software supplied by university[2] |
| | $ 2,000 Cost of equipment supplied by team[1] |
| | $ 4,095 Student edition of PSpice available at          no cost[3] |
| | $ 2,870 Projected expenses |

Example of a Complete Budget for *The Talking Book* Project

How, without experience, will you determine fair numbers? First, check pricing with several retailers of the supplies or equipment you will need to purchase. For instance, the $2200 originally allocated for "RFID transceiver and 10 tags" might represent the price quote obtained from the most expensive of the prospective suppliers. Remember, other suppliers may go out of business or have a backlog of orders from high priority customers, so even though you do not intend to purchase from the most expensive of the prospective suppliers, you might want to plan for this as a contingency. Second, include estimates for all expenses, even if you think they will be negligible. For instance, it would have been difficult to predict the $150 in telephone charges in the final budget. However, by budgeting what would be considered to be a reasonable amount for telecommunications, the team would not be too far over budget, for this line item, at the end of the year.

Also include components that you may be able to obtain for free. Often, electrical engineering senior design teams rely on chip or software evaluation samples provided by industry. If you have not obtained the samples by the time you write the budget, then include the retail price of the components. Again, you cannot guarantee that you will receive the samples on time and you may have to resort to purchasing the components. Similarly, you may depend on a piece of equipment or a service promised by a friend or relative. Again, budget cautiously for any other products or services you hope to get for free, because accidents or misunderstandings often happen (someone may break the equipment or borrow it for another project) and schedules conflict (the cousin is sent away on travel for work just when you need help with testing). If you are hoping to get five items for free, you might decide to include some percentage, for example, 33%, of the retail cost of each of those items in your budget, in anticipation of a 33% chance that the free offer will not materialize. Finally, you will make contingency plans as discussed later.

## 5.4   ECONOMIC VIABILITY ANALYSIS

Let us first consider a budget for a design course. Now that you have considered a few of the simpler forms of a budget, it is time to recall that there are really two aspects to the economic viability of your design project:

- Do you, as a team, have (or are you able to raise) the funds to build a concept-demonstration prototype as a deliverable for senior design?
- Would it make sense, from a business standpoint, to develop this concept into a real product? That is, can you turn a profit when you sell it?

For the first, you need only the engineering development budget as discussed above. For the second you will need to consider the additional costs of marketing and sales, production, support, and administration. Also, you need to think about whether you can sell your product for enough money to turn a profit and when you will begin to make money.

For an industrial project, particularly in a large company, your manager or another administrator decides whether the budget and product are feasible. Your must present a design that is low-cost and that meets customer requirements or design specifications. Even though you will most likely not be responsible for determining economic viability, you should keep these additional costs of your project in mind, since they may affect your design decisions.

Depending on the type of project you are doing, the economic analysis will vary. For instance, if you are developing a consumer product with a large potential market, you will have to do a cost vs. sales analysis as described below. If you are designing a baseball stadium or a component for the space shuttle, your market is limited. In this case you must only determine the design and the one-time construction or manufacturing costs and compare these costs to what the single customer is willing to pay.

An example of a production budget for a typical consumer product is given below. Observe that the costs change over time; a real company will need to focus advertising campaigns at different times during the production cycle and will need to concentrate on manufacturing only after the first orders are secure.

| Expense Stages | Cost |
|---|---|
| Pre-sales costs | |
| Engineering development | $ 550,000 |
| Presales market research | $ 100,000 |
| Advertising and sales staff expenses (prior to first sale) | $ 500,000 |
| Initial manufacturing run (100 units) (prior to first sale) | $    5,000 |
| Subtotal | $1,155,000 |
| Monthly costs from initial sale to 6 months after initial sale | $   50,000 |
| (50 units sold per month) | $   10,000 |
| Ongoing engineering support (@ $50,000 per month) | $   50,000 |
| Ongoing market research (@ $10,000 per month) | $   25,000 |
| Ongoing advertising and sales staff expense | $    5,000 |
| (@ $50,000 per month) | |
| Manufacturing (500 units per month) (@ $25,000 per month) | $  140,000 per month |
| Product support costs (@ $5,000 per month) | subtotal costs initial sale |
| | through 6 months after |
| | initial sale |
| Monthly Costs Beyond 6 Months after Initial Sale | $   10,000 |
| (100 units sold per month) | $    5,000 |
| Ongoing engineering support (@ $10,000 per month) | $   50,000 |
| Ongoing market research (@ $5,000 per month) | $   45,000 |
| Ongoing advertising and sales costs (@ $50,000 per month) | $   10,000 |
| Manufacturing (1,000 units per month) (@ $45,000 per month) | |
| Product support costs (@ $10,000 per month) | $  120,000 per month |
| | subtotal costs beyond |
| | 6 months after initial sale |

Example of Expenses
Incurred during Stages
of the Production Cycle
for *The Voice-Activated
Appliance Controller*

## 5.4.1   Determining Project Costs

How do you determine the numbers? We will discuss some of the issues involved. The first item, engineering production costs is the total of the earlier budget. Recall that in this example, we allocated funding for testing. This sample budget implicitly assumed that we were not proposing to produce a product such as a biomechanical device or a drug, which would require extensive testing, probably on laboratory animals and then on humans. Since this testing cost (in both time and money) might be even larger than the basic design costs, you should budget for it, if appropriate. The next largest cost in the first few months after development will be advertising.

### Advertising Costs

In order to sell a product, you have to make the existence of your product known to prospective customers. If you were working for a large firm, they would hire an advertising agency, probably for much more than is budgeted here. If you are an entrepreneur and are working out of your parents' garage, you would have several options:

- Advertise the product using radio or television.
- Advertise the product in a general-audience print medium, such as *Time* magazine or an airline magazine.
- Advertise the product in a print medium that is likely to be purchased and read specifically by the target customers for your product, such as *Car and Driver* magazine if, for example, your product is a new

type of electronic ignition system that can be installed on an existing car.

- Advertise the product by sending out mailings to prospective customers.
- Sell the product at wholesale to retailers who will advertise it; for example, let Sears or Radio Shack sell your product under their brand name.
- Use the Web to advertise the product, provided you have a way of attracting prospective customers to the product's principal Web site (for example, banner ads).

Advertising costs vary widely. A 30-second advertisement during the Super Bowl costs a lot more than an advertisement in a monthly hobbyists' magazine. For some products, advertising can make up a major portion of the ultimate selling price, because it is very hard to target the advertising to the individuals who will actually buy the product. Your business plan should specify what means you will use to reach your target customers, and the associated cost of advertising should be determined by checking out advertising rates that are published for various advertising methods. It is easy to find out how much it costs to place an advertisement of a specific size and type (for example, number of colors) in a specific magazine.

### Market Research

Market research is an ongoing process in any product cycle. If you have answered a mail or e-mail survey, have completed a questionnaire in a shopping mall, or have served as a Nielsen family for television ratings, you have participated in one of the more obvious types of market research. Typically, you would hire a firm to perform your market research. They would determine whether there is any interest in your product and assist with product pricing. If you worked for a very large company, an in-house division would do this work for you. If you are an entrepreneur, you may save money by doing the market research for yourself. For all cases — barring the self-performed research — we have shown a typical amount required for market research.

The price that can be charged for a product or service depends upon the willingness of the target customers to pay various possible prices and the price being charged by competitors for similar products or services or products or services that serve the same customer need. If there are no competitive products or services, then you can charge whatever price customers are willing to pay. It is usually very difficult to determine this willingness to pay when there are no equivalent products or services already in the marketplace. You can ask people what they would be willing to pay for widget X or service Y, but the answers they provide are notoriously unreliable indicators of what those people will actually be willing to pay when the real product or service is brought to market. If products and services that serve the same customer needs already exist in the marketplace, then it is reasonable to assume that your product or service must be priced at or below the price of the competition. It is never a good idea to assume that you can sell your product at a price that is equal to what it cost for you to make it, plus a desired mark-up. Your competitor's costs may be very different from yours. The most important role of marketing is to attempt to estimate the demand for a product or service as a function of its price, taking the competition into account. Even the most experienced and highly paid marketing experts are often far off the mark when they try to make such estimates. Nevertheless, failing to study and attempt to understand the market for your product or service is asking for trouble.

Obviously, for a design course you will not be able to spend $100,000 or even $10,000 on hiring a market research form. Therefore, you will have to perform an informal market survey. One way to do this is to create a questionnaire that gathers data about the subjects and their willingness to purchase your product. A sample questionnaire for our example product, *The Voice-Activated Appliance Controller,* follows. Note that you should change the questions or the monetary values depending on the product you are developing and your target market.

---▼---

## SAMPLE MARKET SURVEY

### SURVEY FOR *THE VOICE-ACTIVATED APPLIANCE CONTROLLER*

<a description or a picture of the product would appear here>

Are you planning to purchase an appliance controller (circle one) within

3–6 months    6–12 months    1–2 years    2 or more years    never

If you plan to purchase an appliance controller, which features are you most concerned with (circle as many as applicable)

price    color    speed    size

If you plan to purchase an appliance controller, what is the maximum you would pay (circle one)

<$100    $100    $150    $200    $250    $300    $400

Your income (circle one)

<$20,000    $20,000–$35,000    $35,000–$50,000    $50,000–$75,000

$75,000–$100,000    >$100,000

Your age (circle one)

20–25    25–35    35–45    45–65    >65

Comments:

---▲---

You should tap friends, family members, and colleagues to complete the survey. However, when you distribute the questionnaire, make sure that it goes to a representative sample of the target market segment. For instance, if you were redesigning a skateboard, you would not necessarily ask an infirm grandfather whether he would purchase one. Instead, you would target teens and preteens. (Unless, of course, there is some innovation that would make the grandfather feel that this was a necessary purchase for a grandson or granddaughter.) Market research is an art. When you tally the results of your survey, remember that even with your best efforts you have not achieved a complete representation of the market. For instance, you have only obtained the opinions of the subset of people willing to answer surveys. They may or may not have tastes or characteristics that would lead them to purchase appliance controllers. Therefore, it is best to underestimate what people are willing to pay and also underestimate the number of units to be sold.

### Product Support Costs

Generally, if you a produce a product, you will have to support it by such methods as providing support staff or warrantees. Product support staff would range from one person monitoring a phone or Web site part time, to a bank of people who are available around the clock. You should make an estimate of what your product would require. Obviously, the appropriate level of support would vary widely among products designed for different purposes and different customer groups.

## 5.5   ARE WE IN BUSINESS?

Given the above examples, we can determine whether the project is economically feasible and determine which parameters would need to be changed to make the project realistic. For the case of *The Voice-Activated Appliance Controller* or *The Talking Book,* we are building a general consumer product, so we would expect to sell many and eventually turn a profit. In other cases, particularly for industrial projects, you will often have a single customer. Since the goals of these different types of projects are somewhat different, we will analyze them separately.

### 5.5.1   Economic Analysis Method for Consumer Products

Based on the business plan for costs (expenditures) of *The Voice-Activated Appliance Controller* we can create a spreadsheet showing monthly cost and cumulative costs vs. time, starting with the time of the first sale. We can then factor in the offsetting revenues.

For example, if the controllers are sold for $200 each, then the revenues during the first 6 months of sales will be $100,000 per month. The revenues after the first 6 months of sales will be $200,000 per month. If we insert the cumulative revenues in the same spreadsheet as the cumulative costs, then we see that the point where the cumulative revenues equal the cumulative costs is between 23 and 24 months after the initial sale. After that, the product produces an ongoing profit of $80,000 per month. If the controllers are sold for $250 each, then this cumulative cost recovery point moves to between 15 and 16 months, and after that, the product produces an ongoing profit of $130,000 per month.

However, if the controllers are sold for $150 each, then profitability is pushed out beyond the 36-month planning horizon of the spreadsheet. Naturally, the price that can be charged will depend upon the willingness of customers to pay and the prices that competitors charge for an equivalent product. In addition, the number of controllers that can be sold each month will decrease as the selling price is increased, and this will have an impact on the business plan (which currently assumes sales of 1000 controllers per month after the first 6 months). In a real economic analysis, the sensitivity of predictions to changes in the assumptions would be tested by trying out more pessimistic and more optimistic assumptions.

In this example, we assumed that the manufacturing cost of the controllers would drop from $50 per controller to $45 per controller after the first 6 months of production. This is referred to as a "learning curve." As more units are produced, one can assume that costs associated with each unit will drop due to such factors as reduced production time per unit and economies of scale in purchasing components.

### 5.5.2 Economic Analysis for Single-Customer Projects

Some examples of single-customer projects would be large military projects or any engineering design and development project that you have bid on to a company. In this case, there is a clear end user, and typically your design will become the property of your customer. You will be given many constraints, one of which will be a cost constraint. As you perform an economic feasibility analysis on your project, keep in mind that many people will be reviewing the project and that it is your job to convince them that your project is viable. Therefore, you should always consider as many parameters as possible and present them in a clear manner. While you have this contract, you do not have to worry about the advertising and marketing concerns of a multiple-consumer product, nor do you have to worry about undercutting the competition. You do have to come in under budget (or be prepared to lose during the next bidding cycle).

### 5.5.3 Presenting the Economic Analysis

The results of your economic analysis may be displayed in tabular or graphical form. When you display your results, it is important to include at least a brief discussion of the method you used and the assumptions you made when arriving at your figures. Also, if you performed an informal market survey or any other type of analysis, put the results in an appendix, so that readers can review your data and methods.

## 5.6 CONTINGENCY PLANS

In Section 5.3.2 some possible disaster scenarios were mentioned. One that has been encountered by many senior design teams that we know about is the following:

> Just before winter break the team settles on a component that is vital to their project. They call a well-known company, which promises the part in 4 to 6 weeks. This is fine with the team; it will be in the mailbox when they get back to school. Six weeks pass, and it still has not arrived. They call the company, which says "two more weeks." (You can probably predict what is going to

happen.) Two months pass and still no part has come. There are only a few weeks left of senior design, so the team has to scramble to find and implement an alternate design. Finally, the last week of the term, the long-awaited part arrives.

As was mentioned in the previous chapter, you must plan both your budgets and schedule for unfortunate occurrences just as this. It is important that you consider each aspect of your project and that you determine what will happen if a part you need is unavailable, the prototype takes more time to design than you thought it would, or the cost rises to much more than you can afford. We have already given you one technique for dealing with cost overruns, that is, budget for the worst-case scenario (for example, use the most expensive component price that is offered). Second, develop a contingency budget. Perhaps if a vital piece of software becomes unavailable, you can have someone else develop it, or you may have left time to develop it yourself. Obviously, you will have to budget money to pay the subcontractor or time to do your own development. Similarly, students often plan to design a subcomponent of their project themselves and find that other design tasks took up more time than expected. Then they are forced to purchase the component as an unanticipated expense. As you are jointly developing your budget and schedule consider which tasks can be replaced with a purchase, if necessary. Then develop contingency line items in your budget, which you will not include in your primary budget, but which you will briefly discuss in your budget explanation, which is covered in Chapter 6.

A worst-case scenario is that you will have to completely reconsider your project. If you have thoroughly gone through the design process, as described in Chapter 1, you have a number of alternative design solutions. You had eliminated them based on multiple criteria; however, you should be prepared to implement one of these alternative solutions or to return to one of these alternative designs if necessary. To be able to turn to another solution, you must be prepared. Have a second budget and a second schedule (albeit compressed) prepared. You should identify all of the high-risk aspects of your senior design project plan, and you should be thinking about backup strategies from the outset. Remember: In the real world successful people deliver results, not excuses.

## 5.7 SUMMARY

A key idea in budgeting is that budgeting is not simply the development of a table of money spent. Instead, budgeting focuses on the careful evaluation of projects for their economic feasibility and the development of alternatives and contingency plans, in case unanticipated costs arise. Part of the budgeting exercise may be to develop multiple budgets aimed at different audiences and to help you analyze different expenditures in the business cycle. While the budgeting process may initially appear to be difficult, by carefully considering the costs during all phases of the project and having backup strategies in place, you will be able to develop a realistic budget that is acceptable to you and the customer and complete your design project within this allocated budget.

The subject of budgeting is a component of the more general subject of project management. There are numerous books available on this subject, some of which were listed in Chapter 3. These references typically include a section or chapter on budgeting that gives examples of budgets for various types of projects. These sections or chapters also typically include some discussion of best practices in budgeting and common mistakes to be avoided. A few good references are given below.

Archibald, R. (1992). *Managing High Technology Programs and Projects*, 2nd ed. New York: Wiley, Sections 10.4 and 10.15.

Lock, D. (1992). *Project Management*, 5th ed. Hants, U.K.: Gower Publishing, Chapter 4.

Weiss, J. and Wysocki, R. (1992). *5-Phase Project Management*, Reading, MA: Addison-Wesley, pp. 46–50.

★ *Key Points*
*Chapter 6*

- During the design process you will be required to produce several documents, from proposals and abstracts to final reports and user documentation.
- These documents build upon each other, so keeping up with the writing throughout the duration of the project will make your project run smoothly and will save you time.
- Your documents are used for several reasons, such as to convince people to support your project and to disseminate information about your work.
- It is up to you to determine the audience and set the tone of your writing to meet your goals.
- Revision and peer review are two key methods of improving your documents.

# Chapter 6

# DOCUMENTING YOUR DESIGN PROJECT

*Maja Bystrom*

## 6.1 DOCUMENTING A DESIGN PROJECT

### 6.1.1 Before You Begin

While many consider documentation the most tedious part of the design process, there are numerous good reasons for producing well-written, informative, and convincing documents. Putting your thoughts down on paper helps you to organize and refine ideas. Written material, often combined with oral presentations, can persuade an audience to support your effort (financially or otherwise). Finally, writing preserves your work and allows either you or another design team to have a starting point for future work.

You may be required by your company, university, or funding organization to produce a number of documents. These might include:

- A pre-proposal or project abstract
- A proposal
- A project plan
- One or more progress (status) reports
- A test plan
- A report on test results
- A final report
- A user manual (end-user documentation)
- One or more memoranda
- An engineering notebook

Depending on the length or type of the project, not all documents may be needed, and the components of the documents may vary. Typically, the only ones common to every project are an engineering notebook, a proposal, and a final report. However, some documents can serve as the basis for others and can be used to illustrate the extent of the work performed (useful when you are asked for your input on a grade or a performance review) or be used to support an application for a patent.

## 6.1.2 When Do You Start?

Immediately. Starting to write is often a difficult process. However, the best procedure is always to simply begin. It is much easier to start writing and then later return to your drafts to correct mistakes and refine your ideas than to write an entire document at the last minute. Many writers find starting a document easier if they first develop an outline for the document. Following an outline forces you to think about sections you may be missing or have not completely thought through. If you choose not to start with an outline, then it is particularly important as you complete the document to revise and be willing to cut or rewrite sections that do not make sense.

In the most successful projects team members write as they design and implement their project. Starting early and continuing the documentation process is good for two reasons. First, it helps with time management. The more you have done earlier in the project cycle, the more room you will have at the end of the project for document rewrites and schedule slippages. Second, you will have time to put the writing aside and forget what you have done. When you write, you know what you are trying to convey, so it is difficult to step back and get a reader's perspective. If you can put your writing aside for a week or more it is much easier to see how it may be improved.

## 6.1.3 Keeping an Engineering Notebook

Throughout the design process, even as you are composing other documents, each team member should keep an engineering notebook. Not only can these notes be used to build necessary reports and serve as a central location for references, but they can be used to substantiate intellectual property claims. Each page on the notebook should be dated, and all thoughts about or work done on the project should go into it. This material includes, but is certainly not limited to, ideas developed during single or group brainstorming sessions, preliminary and final design sketches, lists of references or possible leads to information about the project, names and contact information for parts suppliers and consultants, along with dates and times of discussions, lists of where information is kept and which team member is responsible for which parts of the project, and summaries of all discussions. The notebook is the property of your employer, and thus the contents and the tone should be formal. See Chapter 1 for more information on format and tone, and Chapter 8 for intellectual property issues.

While it is advisable and often required to keep a physical engineering notebook, some people prefer to work in electronic form. The same note-taking procedures hold; that is, date all of the entries and keep any files well organized. Whether keeping a paper or electronic notebook, each team member must keep electronic copies of all e-mails sent and a list of links to all Web pages with useful information.

### 6.1.4  Save and Back Up

One of the most frustrating experiences during a design project is to have typed several pages of a document and to have all your hard work erased by a power failure or an inadvertent keystroke. To avoid this problem you should save after typing each paragraph and back up after each page. There have been many times when team members have incompatible word processor versions or formats. If one computer crashes unrecoverably (a frequent occurrence), then you must have the document in many forms to be able to reproduce it. Back up onto fixed disk or portable media in the format of the word processor as well as in plain text or other format and be sure to use different file names each time you back up. Finally, print at least once per writing session to have a hard copy of your work.

## 6.2  SEVEN STEPS TO A SUCCESSFUL DOCUMENT

There are seven sequential steps in the writing process that will guarantee a successful document.

- Divide responsibilities.
- Determine tone, style, length, and format.
- Develop an outline.
- Write a draft.
- Revise and proofread.
- Review (peer and other) and revise again.
- Generate the final document.

### 6.2.1  Partition Responsibilities and Agree on Software

Now that you know that you have to start writing as soon as possible, there are a few things to think about as you begin. Unless you are in a large team in industry that has an assigned technical writer, you will most likely be sharing the writing responsibilities with the rest of your team members. As you begin the writing process, your team should agree on the sections each member is to write and select an editor who will perform the final compilation of the material. One strategy for dividing the responsibilities is to have the team members write about the sections of the project with which they are most familiar. For instance, the group member who is in charge of the patent search should document the results of the search. The editor plays a key role in document compilation, with primary responsibilities for both ensuring that all writers complete the assignments, and in collecting and weaving together the disjoint parts. The editor will fine-tune the language in each section to ensure that the document flows well.

Next, you must agree on a word processing program that all team members have access to and feel comfortable using. Before you begin writing a major document, you also must select software for generating tables, images, and figures and check that the software is compatible with the word processing program selected. Do not rely only on the claims of the manufacturer; instead, generate sample figures and images and test the software formats.

### 6.2.2  Determine Overall Look and Feel of Document

Your manuscript has a job to do, namely, to sell your project. Therefore you must be careful about the format, layout, and style; this means determining not only the sections and material to be included but also the tone, the look of the pages, and even the font that you use.

### Style, Audience, and Tone

A good writing style is one of the most difficult skills to learn. One of the best ways to learn an appropriate tone and style is to read and critique many examples of design project documents. If possible, you should obtain copies of reports from previous projects in your organization. Browse through several to get a feel for different writing techniques. You will easily see which ones are interesting and easy to read and which ones are stilted and do not flow well. You will also be able to see what is an acceptable style and format for your organization. Although you should concentrate on developing your own style, you can adopt the more successful patterns and tones from the reports you read.

There are a few basic rules that must be followed to make proposals easy to read: write at the appropriate level and with the appropriate tone for your audience, maintain your style throughout the document, emphasize only necessary sections, make figures and tables clear, use transitions between sections, and make the document interesting as well as informative. As you begin, these goals may seem impossible to reach, but by carefully paying attention to your manuscript and by continually editing, you will be able to attain them.

Before thinking about general layout, you should first consider tone. As you read each example report, you should determine the report's audience and whether your document is intended for the same audience. Who is your audience? Your audience may consist of peers, managers, faculty, external reviewers, competition judges, or venture capitalists. If the document is internal to your organization, it may be read only by a first-level manager who would want a great deal of technical detail and little motivating material. If you are enrolled in a design course, your reviewers may be faculty and peers who are not familiar with your project or the technology you are employing but would be concerned with the motivation behind the project. If there will be a wide variety of readers, you must write for those who are not familiar with your topic but also include enough information to satisfy those who are experts in your field. You may choose to write the body of the document at a level that is suited for the nonexpert and include technical appendices with detail sufficient for the experts. However, you may need to produce two versions of a particular document: one written specifically for a technical audience, which includes many technical details, and one written for a non-technical audience, which gives a broad overview of the technical details but explains the motivation and outcomes thoroughly. As you write each version of the document, you should constantly have the particular audience in mind; this will help to set and maintain the tone of the proposal.

As an additional challenge, your audience may be hostile. They may be ready to dismiss your project because of prejudices against an area or product or may have read many reports on the same day and may be tired. Or they may simply be venture capitalists who need very good reasons to consider your project and then invest capital into it. Therefore, you have to sell your project to your readers. Your primary goal in selecting tone and contents should be to convince your audience that you have done your homework and thought through the project.

There are several ways in which you can do this. The first is to ensure that the project or proposal is complete. As discussed in Chapter 1, you will need to do significant background research and consider all design alternatives. Once you have done a good job of collecting motivating material, you then have to convey how much work you have done. This is not accomplished by deluging your reader with a large amount of information — remember that everyone dreads being faced with a dictionary to read — but by concisely summarizing your work in a manner that is clear to understand.

The second way is to use formal language. Remember that you are not writing to a close friend, so avoid slang and familiarity with your audience. However, you do not want to sound stilted. Although novice writers often associate formality with long words, you should not use words with which you are unfamiliar. You do not need to have a huge vocabulary to be a good writer; you must simply make good use of what you know. One of the few ways to learn whether your document is making a point and sounds good to audience members is to submit it for peer and supervisor review or choose other external unbiased reviewers. Peer review techniques and outcomes are discussed at the end of the chapter.

## Format, Layout, Fonts, Margins, and Other Important Details

Quite often document size and format is dictated by your organization. Even if you can find no formal, written rules, by doing your research on document tone, you would have seen the particular formats used in other projects. However, you should always ask your supervisor, advisor, or mentor whether there is a specific style expected by your organization.

If there is no specific style or if there are only guidelines to document format, then you will have to choose your own format and layout. First, choose the length and allocate a preliminary page count to each section. Once you have selected section length, you should write much more than allocated and then cut back to fit the designated page length. A common fault of inexperienced writers is to misjudge the amount of work needed and to use large fonts or margins to take up space (think back to grade school writing assignments that had page requirements). However, white space, that is, areas on the page that are left blank, is central to ensuring that your readers do not feel overwhelmed. Therefore, there are conventional font and spacing requirements.

Unless otherwise specified, your document should be single to 1½ spaced. Use 11- or 12-point type, and at most 14-point type for section headings. The left and right margins should be between ¾ and 1 inch, while the top and bottom margins should be between 1 and 1½ inches. A smaller font or margin will look cramped and make the document difficult to read. Larger margins or fonts will make your report look like a grade school essay. Page numbering begins with the first page after the title page, which is typically the abstract, executive summary, or first page of the table of contents. All prefatory material is numbered with lowercase roman numerals (for example, i, ii, iii). The body of the report is numbered with Arabic numerals (for example, 1, 2, 3). Page numbers may be placed in either the header or footer and are located either at the center or at the outer corner of the page.

For long reports, each page may have a header. An example of a heading is the chapter or section name, often in bold or italic font, that you see at the top of each page of a text book. For design projects, the left-hand page may contain the title of the report in the header, while the right-hand page may contain the technical report number. The title page naturally does not have any headings. This level of formatting is typically specified by the individual company or department.

To increase the readability of your document, make sure that you spell out all acronyms the first time they are used and limit them unless the acronyms are common and unmistakable. For instance, you should feel free to use "VCR" after spelling it out, that is, refer to it as "videocassette recorder (VCR)." This is particularly important when using an abbreviation such as "ATM," which is commonly used for "automated teller machine" as well as for "asynchronous transfer mode." Also, do not use Latin abbreviations or words. They are often used incorrectly, and some readers may not be familiar with them. English phrases can easily be substituted ("that is" for "i.e.", "for example" for "e.g.") in place of these abbreviations to make your writing clear.

Make sure that you spell check before handing the document to an outside reviewer. It is also good manners to spell check even when giving text to a team member. It is best if you make a habit of using a spellchecker and grammar checker while you are writing, and at minimum before you make a periodic backup of your work.

### Emphasizing and Delimiting Ideas

You may use italics, boldface type, or bulleted or numbered lists to emphasize the most important points of your writing. However, writers often overuse emphases so much that the extra delimination becomes distracting. As an example, in your discussion of patents, you may use bullet points to break up text. Clearly define major sections so that your readers can find them. Unless you have an overwhelming number of items to discuss, do not use lists or small sections.

Below are samples of poor and better uses of emphasis. Unless specified in document guidelines, it is generally not necessary to put each component into a separate section. The second example uses boldface only for section headings and uses no underlining or other emphasis beyond italicizing the type of microcontroller. When combined with a clear line drawing (the associated drawing is given in the Appendix in the draft final report for *The Talking Book*), this example is concise and yet informative. Additional information about the choice of the microcontroller can be placed in this section or in a section on design constraints.

---
▼
---

## POORLY WRITTEN

### 3.2  Design Overview

The book has *five* components:
1. on/off switch
2. play/record switch
3. information storage device
4. radio frequency identification device
5. microcontroller

### 3.2.1  On/off switch

This is a toggle switch located on the back of the book. It controls power to the reader and the microcontroller.

### 3.2.2 Microcontroller

We have selected the <u>National Semiconductor COP8SGR7</u> microcontroller, since we must use a UART. The features of the microcontroller are:

- *integrated* UART
- low-power 3 V requirement

### 3.2.3 Play/record switch

This is a switch located on the back of the book and interfaced with the microcontroller. It allows the user to switch between two settings, play and record.

## IMPROVED VERSION

## 3.2   Design Overview

The book has five primary components:

1. on/off switch
2. play/record switch
3. information storage device
4. radio frequency identification (RFID) device
5. microcontroller

These five components are shown in Fig. 1. The book will have two external, user-controlled switches, the play/record and the on/off switch. Both are toggle switches that are linked to the low-power *National Semiconductor COP8SGR7* microcontroller. Information about the current page is transmitted from the RFID to the microcontroller through an RS-232 connection.

───────────────────▲───────────────────

## References

Citing published works in both electronic and paper form is a valuable tool for ensuring that you have thoroughly investigated your problem and for convincing your audience of the same. While it is generally easy to find information online, the accuracy of this information is not guaranteed. For the most part, information on the Web is not peer-reviewed or even reviewed by an editor, so there is no reason for this information to be accurate. Furthermore, unless stored on the Web site of a standards body or professional organization, online works are fleeting, and it is unlikely that future readers will be able to find your electronic references. Therefore, it is a good rule of thumb that all undergraduate reports should reference at least four printed works; in general, reports should include as many printed references as necessary to support the claims of the work. Any facts obtained online should be confirmed by a reliable printed outlet.

As you find possible references, you should record the reference's information. When you determine the final selection of references you will use, these must be listed in the references section. Each reference is numbered and listed in the order in which it is cited in the text. When citing a work in the body of the document, place the citation number in square brackets (for example, [1]).

A commonly used form for bibliographic references in engineering publications is:

**Journal Article:** J. Smith and R. Jones, "Title of the Paper," *Journal Title*, vol. 1, no. 1, pp. 33–51, June 1996.

**Article in Conference Proceedings:** J. Smith and R. Jones, "Title of the Conference Article," *Conference Name* (City, State/Country), vol. 1, no. 1, pp. 15–20, May 1996.

**Book:** J. Smith, R. Jones, and G. Howe, *The Book Title*, New York: Publisher Name, 1981.

**Edited Book:** W. Meyer and T. Thomson (eds.), *Edited Book Title*, New York: Publisher Name, 1981.

**Patent:** "The patent name," US Pat. X,XXX,XXX, L. Jones and D. Roberts.

**Online Reference:** Title of page or author/editor. (Date of posting or "No date"). Title of site or material accessed. http://www.siteaddress.com. [Date of access].

### Figures, Tables, Flowcharts, and Images

Figures, tables, and graphics can be some of the most useful tools in conveying information. However, if they are poorly created, labeled, or placed, they can be intolerably frustrating to your reader and can damage your project if they cause the audience to draw incorrect conclusions. Each graphic should add some value to the text. If you find you are putting a figure in to take up space, then it is likely to be useless.

Each figure and graphic should have a caption that summarizes the ideas you wish to convey, and all wording must be in a font large enough to read without straining. If it is cluttered it will be difficult for your reader to understand, and you should consider breaking it into two or more figures. Clearly label all lines in figures and rows or columns in tables. The following figures show samples of poor and better data plots. Figure 6.1 has no axis labels and nondescriptive legends, and it is difficult to determine the difference between the lines. Figure 6.2 has one axis on a logarithmic scale, which emphasizes the differences between the data points. It is clear by the plotting style that there are eight distinct points in the data set. The axes are clearly labeled with units and the legend is slightly clearer.

System diagrams and flowcharts are vital for explaining complicated systems and very useful in assisting explanation of even simple systems. However, you must be careful to ensure that they are readable, that is, that each diagram is not too complicated to be very easily understood. One method of doing this is to use a high-level diagram as shown in Figure 6.3. Only the three major system components and primary interconnections are shown. It is clear that there is at least some input and output (speaker, microphone, switches), but the level and operation have been abstracted. Second-level diagrams could be made of the LED circuits or speaker and microphone amplifiers and filters, if necessary for presentation.

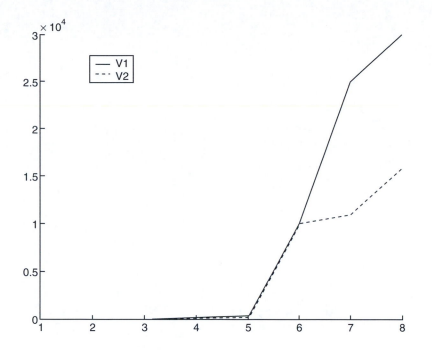

**FIGURE 6.1**
Example of a poor figure.

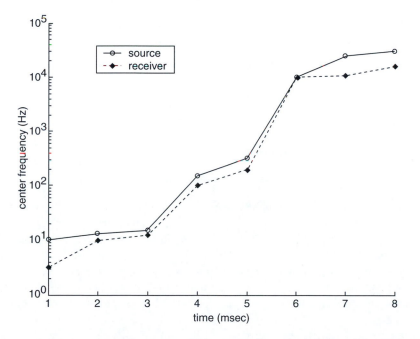

**FIGURE 6.2**
Example of a better figure.

 Pfeiffer's book (2000) contains an excellent chapter on the use and layout of graphics. You can also find numerous references in technical writing books in your library. We encourage you to consult these references and also consult your advisor and peers on how to best portray the information. Ask for honest opinions about whether the graphics are readable and helpful, or whether they are confusing, and then modify your figures accordingly. Your audience will be grateful when they see that you have taken care in generating useful graphics.

 As a final note, photographs can also be useful in illustrating your design. For instance, if you wish to illustrate the size and shape of a final packaged product, you might photograph it beside a ruler. In any case you must ensure

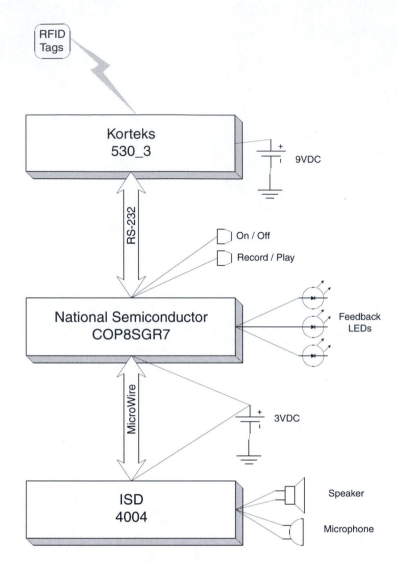

**FIGURE 6.3**
Sample system diagram.

that the object is photographed against a contrasting background, that the resulting image is of good quality, and that the product is in focus and clearly displayed. It is also possible to edit photographs to add labels on components or to point out particular features. However, the components and labels should be clearly visible and the purpose of the labels apparent. Finally, the photograph should reproduce well in black and white, because color reproduction is not always available.

### 6.2.3   Develop an Outline

Before you begin writing a document, even only a one-page document, it is good practice to generate an outline with topics, preliminary page counts, and a tentative list of figures and tables. The outline helps you focus on what is important in your document. For instance, should you allocate the same amount of space to background and motivational material as to the proposed solution? The answer to this question naturally depends on the document and the audience. Having a rough idea of section length will also give you and your teammates an idea of the level of detail to use when writing.

Determine your overall page count either from limitations or suggestions given by an advisor or manager or by looking at similar documents produced by your peers or others within your university or company. List the sections you will need to include and allocate space according to comparable sections in a sample document. If none is available, then you will simply have to use common sense. Ask yourself: What are the most important ideas to convey and where should they be located? You will then allocate most of your space to those sections, while not neglecting others. Recall that the two most important sections are the introduction and the summary. Your introduction provides your reader's first impression of the project, while your summary provides the last. You need to allocate as much space as necessary to make a good impression. You certainly do not want to run out of space at the end of a 20-page document and try to conclude and reinforce your ideas in a short paragraph. A rule of thumb is to dedicate one-tenth of the total page count (excluding appendices) to both the introduction and summary.

As a team, you should agree on figures and tables and the general shape and contents of these. For instance, ask yourselves whether you will need flowcharts to explain a process or a block diagram to show system connections. Will you need graphs to explain pricing trends or consumer interest in a product? Agree on what will be included in your budget or material tables. However, even if you cannot agree on all of these contents immediately, then begin anyway and discuss as the document evolves; you can always remove extraneous text and figures.

An example of a possible outline for the *Coreware* project 10-page proposal is given below. Note that the initial outline for this document is very brief, with just a few placeholders to remind the team of the contents of the different sections and initial decisions on figure and table placement. Since this is a proposal, the team has to sell the idea and convince the readers that they have a feasible solution in mind, so most of the document space is devoted to motivating the problem and discussing the preliminary ideas for the solution.

▼

## SAMPLE OUTLINE FOR *COREWARE* PROPOSAL

| Section | Approximate Page Count |
|---|---|
| Introduction | 2 |
| Background, Motivation, Problem Statement | |
| Solution Constraints | |
| Figure 1. Network Interface | |
| Overview of Proposed Solution | 4.5 |
| Simulation, HDL description | |
| Alternative solutions | |
| Project Schedule | 1.5 |
| Figure 2. Gantt chart | |
| Project Budget | 0.5 |
| Table 1. Budget | |
| Overview of Impacts | 0.5 |
| Summary and Conclusions | 1 |

▲

As a contrasting example, consider the next example, which might represent a section of a proposal for developing a videocassette recorder (VCR). Due to the complexity of the system, this document might be long (perhaps 50 pages). In this case more planning is required and the outline is much more detailed.

▼

▲

### 6.2.4  Write a Draft

Sit down, follow your outline, and write your draft. For short documents (for instance, fewer than 10 pages) you might plan to write in one (often long) sitting. For longer documents, it is best to write drafts of sections independently, but keep a notepad at hand should inspiration about a section come while you are working on another. It is best just to start writing, and then worry about wordsmithing later. The first draft you write will not be the one you submit, so it does not have to be polished.

### 6.2.5  Proofread and Revise

It is often said that an author is never finished, that a document can be written and revised indefinitely. While this is certainly true, what this means to you is that you are not finished once you put words in each of the document's sections. Recall that your job is to sell your work, that is, to make a clear and compelling argument for it. This is rarely achieved with the first draft. Instead, you should make time to set aside and then revisit your work periodically.

Revising or editing your work can be one of the most tedious, but it is certainly one of the most important, processes in writing. Generally, it is easier to revise your document after you have set it aside for a period of time so that you have a chance to clear your mind and step back from the writing. Of course, with a busy schedule and the limited amount of time in a design project, this is easier said than done. The best methods of ensuring that you have time to revise are to begin writing during the project design phase and to allocate time in your timeline specifically for editing. Even a few days' break should be sufficient time to give you some perspective on the document.

Each time you read your work in preparation to revise it, you should ask yourself whether it is conveying exactly what you want it to convey and whether it has a tone appropriate for the intended audience. If it does not, you must rewrite or rearrange sections. It is often difficult to remove sections on which you spent a great deal of time, or to convince team members that their work needs editing, but your document often will benefit enormously from both. If you are responsible for writing a section and you feel that it does not fit in with the surrounding material or does not sound the way you wish it to ask your team members for suggestions.

Even small errors can easily convey the idea that you do not care about your work enough to pay attention to details. Spell check carefully, paying extra attention to common errors such as transposed homonyms or misplaced apostrophes. A writer's handbook such as that by Alred et al. (2003) can be very helpful for confirming simple grammatical rules.

### 6.2.6  Peer Review and Final Revision

Peer reviews are also vital to the writing process, mainly since they provide assistance in the revising process. You should have at least three reviewers who are not familiar with the project read and critique each document, and you should take these reviews to heart. If your reviewers are confused or are not convinced by your proposal, then you have not done your job adequately. If the reviewers cannot understand the steps in your user documentation, then it must be rewritten. If you have not conveyed the purpose of your project and your design steps, then you are not meeting your audience's needs, no matter how much time and effort you have put into the writing process. Even if you feel you have addressed or even overemphasized topics, your audience may need reminders of your goals throughout the document. Peer review is a very good way to gauge the mood of a real audience. Additionally, reviewers may capture errors in spelling and grammar not caught by an automatic checker and may have useful comments about style. This is especially important when the language of the document is not your first language.

Occasionally you may receive conflicting advice from reviewers. This is natural since all reviewers have their own perspectives. If comments from the reviewers are extremely detailed such as "you have discussed topic X too much" and "you need to talk about this aspect of topic X," then it is difficult to satisfy both readers. The only approach is to try to write more tersely and include the additional information, or move the detailed discussion of topic X from the body of the work to an appendix, include as much as you like, and simply leave an overview of the topic in the body. If you receive conflicting information such as "this is the best proposal I have ever seen" and "I do not think this is a worthwhile avenue of pursuit" then you should go back to the reviewers and ask for more details. Find out why they did or did not like the proposal. Perhaps they did not understand your goals, or perhaps they have seen someone try the project before and know it will not work. Also, even when reviewers have given you extremely favorable reviews, if they have not included many details, then return to them and ask what in particular they liked about your document. Remember, you want to repeat your successful techniques in your next document.

### 6.2.7  Preparing Your Final Document

*Printing or Copying*

If not otherwise specified, you should print double-sided pages on clean white paper. The first page of the body of the document will be on the right-hand side when open (see page 1 of this book as an example). For design class reports, home printing is usually acceptable, as long as your printer is high quality and figures and images reproduce well. For short industry reports, a department copier may be used, again, as long as the results are high quality. In either case you must check ahead of time that the copier is working well, that is, not producing spots, streaks, or faded areas, and that at least one backup toner cartridge is available.

For class assignments you might also consider having copies professionally printed or reproduced at a university or local copy shop. Be sure to either schedule a time to have this done, since many of your fellow classmates will have the same assignment due at the same time, and be sure to have a backup shop in mind in case the primary one is too crowded or must close for some reason.

For industry documents, there often is a department or building copy shop available. Again, you probably will be able to schedule a time for copying. Make sure that it is well in advance of the due date, so that you can locate another copier in case of emergencies. Also, the shop may require a department charge number, so be prepared and have the copy shop account set up well in advance.

### *Document Presentation*

Be aware of the format and any procedures required for submission of final documents. If you are preparing a report for a class, the requirement may be as simple as submitting several copies printed on white paper and stapled in the upper left corner. This should be your default method of submission of papers for peer review if the document is short. If your are given no other guidance or constraints, then you should affix your document in a clear binding that you can obtain from bookstores, paper stores, or copy shops. This gives it a more formal and professional look than simply copying and stapling.

However, if you have long report (more than 50 pages) you should have it professionally bound (a service available at copy shops), or, at minimum, put into a three-ring binder. For very long reports, you should indicate section boundaries with tabbed pages, colored paper, or other markers in order to make it easier for your reader to locate sections.

If you are given submission requirements, such as paper stock or binding format, you naturally must follow them. Be sure to obtain any necessary binding supplies well in advance or schedule time with the university or company binding office so that you are certain the document will be completed sufficiently in advance of the due date.

## 6.3   DOCUMENTS AND THEIR COMPONENTS

You may be required to produce a variety of documents. In this section the most common documents and their components are discussed and samples of some of these documents are given.

### 6.3.1   Memorandum

A memorandum is typically a short note that is used (among other reasons) to formally notify people of important events, to confirm decisions, or to communicate a brief idea, such as the weekly progress on a project. Memorandums (memos) can also serve as longer documents such as internal reports, but are typically not distributed outside a department or company.

SAMPLE MEMORANDUM ON A WEEKLY STATUS UPDATE

**Memorandum**

Date: March 19, 2004

To: Dr. Edith Jameson, Associate Director, Eng. Dept.
    Dr. Lawrence Smith, Senior Staff

From: David Jones

Regarding: Status of Design Project #16 for the week of March 15, 2004

On March 18, 2004 the design team met and completed the re-design of the user input module. This completion brought the team ahead of schedule. It is now estimated that the initial testing phase will begin the week of March 29.

## 6.3.2    Abstract or Pre-Proposal

### *Purpose of an Abstract*

An abstract is a one-paragraph to multipage document that will serve as a pre-proposal, a stand-alone summary of your work, or a component of a proposal or report. If you are not writing an executive summary, the abstract will serve as the first page in your proposal and your interim and final reports. The primary reason to include an abstract is that it serves as the introduction to your work — the frontispiece — and will be what most people will remember about your project. Good first impressions are vital. If the abstract is well written it will capture your readers' attention and inspire further reading. If it is poorly written, boring, and uninformative, it is likely to alienate the readers and prejudice them against your project.

Another use of an abstract will be to obtain preliminary project approval from a manager or oversight committee. In both academia and industry you will need to have approval from someone to start the project. For the same reasons given above, you will want your abstract to be concise and interesting. It should summarize the idea and provide at least a sentence about motivation for the project; that is, it must contain high-level details about the project, as well as an overview of why it is useful or necessary. One major failing of many abstracts is that they provide too much motivation and not enough detail; that is, they are too vague. With only one well-written sentence you can convince your reader that there is motivation for solving the problem, but you must also convey the fact that you have at least one approach in mind and are aware of the potential major impacts and obstacles.

If you are a student, your reason for writing an abstract may be to use it as a pre-proposal to find an advisor for the project. A busy faculty member or industry group leader will be more likely to want to support you if you have a well-written abstract with clearly presented ideas. It shows that you have put effort into thinking about your design, can articulate the reasons why your project is feasible, and have a direction for the design or solution of the problem. Providing potential supporters or advisors with copies of an abstract before a meeting, rather than having them hear the idea for the first time as you try to describe a project, works for you in two ways. First, it gives the reader a chance to mull over ideas and develop a list of questions. Second, it ensures that you have not forgotten any important aspects of the project (which is very easy to do when you are trying to describe a complex process verbally). Finally, the abstract can serve as a one-page summary of the project for public relations purposes, perhaps printed in a department annual report or used with a poster presentation or distributed at meetings and conferences.

Your abstract length will vary with purpose and requirements. However, whether your abstract is an introduction to a document, a pre-proposal, or a standalone summary of your project, it still must have a number of the same elements. You must remind or introduce the reader of the motivation behind the project (the *why* of the project) and cover the approach to the solution (the *how* of the project) and the major obstacles to or impacts of the solution. If there are any unusual aspects such as the participants (the *who*), the place the work will be performed (the *where*), or any special equipment or supplies needed or processes followed (the *what*), then these should be mentioned. For instance, if you will complete some work in an industrial site, perform experiments in the field, or send a component out for fabrication, this must be noted. If your abstract is serving as the frontispiece to a document, it must also reflect the content and purpose of the document. For instance, if you are writing an interim report, the abstract should cover the state of the project, that is, where you are on the time line, and any major changes in approach or new hurdles to overcome.

### Abstract Style

Below are samples of poor and better abstracts for a multidisciplinary project. These are examples of brief abstracts that might serve as introductions to a proposal or as pre-proposals; that is, these abstracts simply outline the proposed project idea.

──────────────────▼──────────────────

## SAMPLES OF POOR AND BETTER ABSTRACTS

### POOR ABSTRACT

It is widely known that a parent's reading to a child can improve the child's reading and writing skills and encourage the child's intellectual development. Despite this well-recognized need for adults taking active roles in encouraging children to read, very few products are available that creatively utilize state-of-the-art electronic components to provide efficient, cost effective, and entertaining reading aids. *The Talking Book* will harness the power of digital circuitry and a radio frequency identification (RFID) sensor system to record a parent's voice as he or she reads the book. The child will then hear the story being read back in the parent's voice as he or she pages through the book. *The Talking Book* will allow the concerned parent to encourage a child to read even when he or she cannot be physically present. *The Talking Book* will revolutionize the reading experience for every child through advanced technology with a humanistic touch.

It is widely known that a parent's reading to a child can improve the child's reading and writing skills and encourage the child's intellectual development. Despite the well-recognized need for emphasis on reading in early childhood, there are few products available to assist busy adults in reading to children. *The Talking Book* will be an attachment to a standard children's book and will be light enough and simple enough to be operated by children. It will harness the power of digital circuitry, a built-in microphone and speaker, and a radio frequency identification (RFID) sensor system, to record up to 8 minutes of a parent's voice as he or she reads the book. The child can switch the book on and then turn the pages of the book to hear the story read in the familiar voice as many times as desired. The RFID sensor will determine when pages are turned to keep the reading synchronized with the open page. The final product will be expandable so that additional memory can be added to record longer books, and so that it can be used to record multiple books. At a final cost of under $50 per reader, *The Talking Book* will be a product that is affordable to many and can help encourage a child to read books.

The first abstract is written mainly in a passive voice. Observe how much stronger and assertive the second abstract appears. Little detail about the project is provided in the first abstract, and a sweeping generalization is made about revolutionizing the reading experience for all children. The only concrete point is that RFID will be used. In the second abstract, the authors present more substantive goals and design ideas. From this abstract we learn such details as the final product will be hand-held; it will cost under $50; and it will consist of multiple specific components.

The following example illustrates the differences between abstracts for proposals and status reports. The stand-alone abstract gives some background information on the project as well as some directions (this project is developed in hardware description language (HDL), and is meant to be implemented on an field programmable gate array (FPGA)).

## DIFFERENCE BETWEEN AN ABSTRACT AS A PRE–PROPOSAL AND AN ABSTRACT IN A FINAL REPORT

**Pre-Proposal or Proposal Abstract**

Currently we are at a crossroads in the growth of the Internet. Internet Protocol v. 4 (IPv4), the dominant protocol format for the network layer, is beginning to reach its operational limit. IPv6, or the next generation Internet Protocol, will alleviate the congestion associated with IPv4, i.e., small address space, by expanding the address space and allow for prioritization of data streams depending upon their content.

We propose to develop a networking technology that would bridge IPv4 to IPv6 packets. The goal is a low-cost, extremely high-throughput implementation of a bridging technology. This product would be designed to be implemented on an FPGA. This solution would be developed in HDL, with the final product being a synthesized core, intended to be a piece of intellectual property that would be easily adaptable to other companies' designs, allowing for integration into future products.

**Final Report Abstract**

The explosive growth of the Internet is rapidly exhausting the current address space allocated by Internet Protocol version 4 (IPv4). This problem, coupled with recent "denial of service" attacks, has demonstrated a critical need for a new Internet Protocol. The Internet Engineering Task Force (IETF), in order to rectify the problems associated with IPv4, has released Internet Protocol version 6 (IPv6). Due to the radical changes deemed necessary by the designers, the two protocols are not interoperable. The size of the Internet and the existence of legacy software inhibit a speedy migration from IPv4 to IPv6. In fact, this transition will take years, if not decades, to complete. The continued growth of the Internet hinges upon the adoption of IPv6. So as not to segregate the Internet into two separate islands, one based on IPv6 and the other on IPv4, there is a strong desire for a bridge to span the divide.

This report describes how the design team was able to develop a network level device that will allow for communication between the IPv4 and IPv6 networks. This product was developed using recent advances in digital hardware design, allowing for a high speed, low cost product. The solution described in this report is delivered in the form of an Intellectual Property block, or coreware. This design may be licensed to hardware production customers to embed in their network level devices. The development of this bridge between the two network protocols will ease the transition from IPv4 to IPv6. Ultimately, this coreware solution would facilitate an accelerated migration to an Internet that provides improved security, quality of service, and ease of configuration.

▬▬▬▬▬▬▬▬▬▬▬▬▬▬▬▬▬▬▬▲▬▬▬▬▬▬▬▬▬▬▬▬▬▬▬▬▬▬▬

The abstract for the final report describes the contents of the document and gives an overview of the product. To improve the second abstract the team could provide more details about the final product, such as that is was developed in HDL, is meant to be implemented on an FPGA, and simulations show that with the architecture, headers can be translated within 10 clock cycles.

### 6.3.3 Executive Summary

An executive summary is a brief summary of the proposal or report intended to provide a busy reader with an overview of your work. Often managers will not have the time to understand all of the details and nuances of the projects they oversee, or venture capitalists will not have the time to completely read every proposal that comes across their desks. Therefore, they need a brief introduction to the work with sufficient details on the budget and economic analysis to convince them that the project is worthwhile and that you have the knowledge and skills to complete it. Typically, executive summaries range in length from one page for reports of 10 or fewer pages to 10 pages for reports with lengths on the order of hundreds of pages. They are similar to abstracts but often contain more information about project duration and cost and fewer details about implementation.

Begin the summary with a concise statement (two to three sentences for a short report) of the problem to be solved. Often students are overly ambitious in their introduction; they feel that to catch a reader's attention or to motivate the project, they need to state how they are going to cure the world's problems. Poorly written summaries typically begin with broad statements such as "One million people die each year of disease X; by a redesign of component Y we can save these people." This is a noble goal, but the readers realize that you would be unlikely to achieve it (otherwise, except in very rare cases, it would have already been done). A more reasonable statement would be "One million people rely on device Y to combat their disease. By redesigning component

Y we can decrease its cost by 8% and increase the component lifetime by 50%." The reader automatically draws the conclusion that this project has potential for significant social impact, without the bluntness of the first statement. The readers also have hard figures presented to them in the first paragraph; savings in price and increases in efficiency tend to draw readers further into the document.

After you have subtly captured the reader's attention, it is then time to summarize the major points of the document. Describe the problem statement in slightly more detail. For instance, is component Y unaffordable for many people? Who else would benefit from your design? In one or two sentences, summarize your economic analysis.

Next, describe the significant design decisions. Is this a hardware or software solution? Are you sintering vs. forging? Are you using concrete vs. brick? What are the major decision criteria you used? Again, elaborate explanation is not necessary. It often suffices to say, "The code will be written in Java due to the expertise of the project members in this programming language and to ensure that the code will be portable to the three platforms in wide use at the company."

Provide some information about your design steps. It is likely that your reader will not know anything about your field. Therefore, you do not need much detail, but you do want to convey to the reader that you have thoroughly thought through the design process. If significant equipment, funding, or consultant time will be required to complete the project, then you must mention this and briefly discuss contingency plans.

### 6.3.4 Proposals and Reports

Every proposal or status report has some of the following components:

- Title page
- Executive summary
- Abstract
- Table of contents
- List of figures
- List of tables
- List of abbreviations
- Introduction
- Body
- Summary
- Glossary
- Appendices

We will cover each of these components individually and then discuss how to weave them together into a single document.

#### *Document Components*

*Title Page*

A sample title page from the *Coreware* project is given in the Appendix, while a sample page from an industrial project is given in the example below. In general a title page will contain the project title, the authors, and the employer or entity (such as the customer, team advisor, or course coordinator) for whom the work is being performed. Each company or university will have a different required format; some may assign report numbers, some may want only the project title and date on the first page. Check with your supervisor, advisor, or the course coordinator for the correct format. Make the project title descriptive, but keep it brief. If it sounds too long, it probably is. Alternative titles for the example could have been "A VCR," which is too brief and not descriptive, or "A Low-Cost Solid State Plasma-Screen Handheld Color Video Recorder and Player," which has more information than needed.

▼

## SAMPLE TITLE OR COVER PAGE

Proposal No. 129-13-682

February 23, 2003

**Portable Color Video Recoder/Player**

Submitted to:

John Smith
Associate Director, Research and Development
IMTB Labs, Inc.

Submitted by:

Mary Jacobs                     Bill Jones
Senior Research Staff           Staff
Software Research Division      Software Research Division
IMTB Labs, Inc.                 IMTB Labs, Inc.

▲

## Table of Contents

A possible Table of Contents for the *Coreware* project is given in the Appendix. The level of detail of the Table of Contents, that is, the level of subsection you should list, varies on the basis of document length and requirements. For short documents, as shown in this example, you should not go beyond a few levels.

▼

## TABLE OF CONTENTS

▲

The List of Figures, List of Tables, and List of Abbreviations are not necessary unless there is a large number of each of these. The lists of figures and tables generally take the form given below, with the caption given along with the figure or table number and the page number. You may choose to put the List of Abbreviations in an appendix.

▼

### LIST OF FIGURES

▲

## Executive Summary

An executive summary is a high-level overview of the project. It is aimed specifically at managers or funding organizations and must have some details regarding project purpose, implementation, cost, performance, etc.

## Abstract

The abstract is a brief introduction to the document. It should contain a very brief overview of the project, but then contain details about the document such as the purpose of the document and what will be shown in the document.

## Proposal or Report Body

A number of questions should be answered in the report or proposal body. Sample questions are listed below:

1. What is the problem to be solved or the need to be addressed?
2. What is state-of-the-art and what related work is there in the area?
3. Will you build on previous work or unique principles?
4. What is your solution to the problem?
5. What additional research must be performed and at what phase of your project?
6. What are alternative solutions or alternative components to your solution?
7. Are there any key components on which your solution rests?
8. What will you do if these components are not available or if you are unable to design or build them; that is, what is your contingency plan?
9. Do you have the necessary skills, such as programming experience, to be able to complete the project, and, if not, how will you obtain the required skills?
10. Where will you do the work, for example, at your university or at an industry location?
11. Will you perform experiments and will you need special facilities or assistance for these experiments?

12. How will you test the solution?
13. What criteria will you use to determine that you have an acceptable solution to the problem?
14. What will it cost; that is, is your project economically feasible?
15. What are the environmental and social impacts?

Depending on the type of project, not all of these questions must necessarily be addressed, but you should carefully consider them anyway. The questions do not have to be answered in sequence and should be answered in prose form; that is, the answers should be carefully woven into a few pages of text. One mistake authors make when presented with a list of questions is to present the questions and answers in sequence. This makes for very dry reading, and it appears to the reader that the authors considered only what was required and no more. You should make it clear that you have considered questions such as this, but also have considered the unique aspects of your project. These answers can be arranged in a multitude of ways. A typical section division is as follows. Note that these are not necessarily the appropriate section headers but show a possible logical flow of information.

*Introduction or Problem Statement* This section is intended to both draw the readers into your document and to give them a general idea of the problem you are addressing and why it should be addressed. Typically, this is an expanded version of your abstract, with an overview of background material and with cited works to support the motivation for the work. A vital component is a list or table of design constraints and customer specifications.

*Patent, Product, and Literature Review* The overview of patents, related products, and the literature is generally a part of the "Problem Statement" section, because it forms the motivation for your project. However, you can also place this in a separate section.

If your project will require the use of components that are patented, or if the work you are proposing is related to current patents, then you must refer to these patents. If there are many patents that are related to your work, it is a good idea to summarize primary techniques or ideas and then include a complete list and discussion in an appendix. Similarly, there may be products, buildings, or machines that have been previously designed or are currently in existence and that are related to your project. You should refer to each of these in the body of the document and include a brief overview of the related work and a discussion of how your project differs from this previous work. If the patent, idea, or product is extremely complex, you should also devote an appendix to a thorough description of the related work. A good rule of thumb is that for a 10-page paper, one page should be devoted to discussion of the background material. In the rare case when there is no prior work, you should at least mention that you have performed a search for patents or related products and were not able find any.

------------------------▼------------------------

**SAMPLE OVERVIEW OF PATENTS RELATED TO *THE TALKING BOOK***

In the past, parents and teachers have given children storybooks along with audiocassette tapes or records which would assist the child in reading the book and take the place of the parent's reading. Patents have been filed and products have been sold using cassette tape devices inside books and attached to books. While a cassette can be controlled and indexed, it is hardly an easy-to-use, long-lasting solution to the problem. Other patents have been filed incorporating digital audio devices, some even with devices to determine when a page has been turned in order to play the audio relevant to that page. However, the existing patents call for using "switches" and "inserts" (US5374195) as well as a "capacitive sensor" (US5569868); even "depressible user response buttons" (US4997374). We find that the existing patents are not suitable for application where a child is involved, and operation of the device must be simple. Significant technological improvements have been made in the fields of audio digital signal processing and electronic component miniaturization such that bulky external recording and playback devices are no longer necessary.

▲

If, however, the goal of your project is to improve upon an existing design, then you must discuss this former design in depth in the body of the document. Include the design trade-offs, economic analysis, and other pertinent information about this existing design and then expand further upon these topics in an appendix. For example, the *Coreware* project implemented an existing design. A section of the text from the *Coreware* final report given below discusses the specifications for the design as given by an industry standards body and presents some prior work. This example refers to the standards documents that contain the specifications for the two header types to be translated (references [3] and [4]). References [8] and [9] refer to competing solutions, and detailed flowcharts were placed in appendices to explain the flow of the translation. An area where this background work might be improved is in justification of the conclusion that "neither document provides an acceptable solution."

▼

## SAMPLE DISCUSSION OF PRIOR WORK

We propose to implement the IPv4 to IPv6 Bridge in hardware by utilizing HDL, specifically VHDL (Very High Speed Integrated Circuit Hardware Description Language). Improvements in HDL allow for complex algorithms that were originally constrained to software implementations to be placed completely in the hardware domain. In recent history, high-level HDL design tools have become the accepted methodology for logic designers [7]. The remainder of this section will be devoted to the research and design of the network level device.

▲

Research was first undertaken to determine the actual definitions of what constituted an IPv4 and IPv6 packet. Most of these answers were found in RFCs. These documents form the standards upon which the vast majority of Internet protocols are based. The RFCs of particular interest were RFC-791 and RFC-2460, on IPv4 and IPv6, respectively [3, 4]. Upon gaining a firm understanding of the two packet formats, the design team was able to develop a pipelined architecture to translate between the two protocols.

A translation scheme of this nature was described in RFC-2765, entitled "Stateless IP/ICMP Translation Algorithm (SIIT)" [8]. This RFC was then implemented in the form of a Windows NT device driver, by a team at the University of Washington [9]. Using RFC-2765 and the report from the University of Washington team, pipelined algorithms have been developed to handle the four different possibilities that may occur while translating between IPv4 and IPv6 [8, 9]. We have developed flow charts to detail the translation scheme for each of these header fields found in Appendices K–N. Unfortunately, neither document provides an acceptable solution for translating the address fields contained within the packets.

▲

Finally, including a literature review, that is, a summary of existing print and electronic documents that are related to your work, is vital. A detailed literature review ensures not only that you are knowledgeable about your design, but that your idea is unique or has unique features, and it convinces readers that you have done a thorough job in evaluating the trade-offs and comparing your project to existing work. A literature review is the discussion of any literature relating to your project. It is similar to a patent or product review, but is focused on printed or electronic documents. What you should search for is any material such as articles, statistics, or opinions that will support your design decisions. Articles showing that your selected database or programming language will be supported on a number of platforms in the future are relevant. You should also look for documents that can serve as motivating material. Experiments showing that your project is needed, as well as statistics or quotations from expert sources, work well to justify ideas.

An example of a literature review from the *Coreware* project is given below. This segment sets the stage for the problem to be addressed and begins to justify the need for efficient format conversion methods.

▼

### SAMPLE LITERATURE REVIEW

In recent years, the growth of the Internet has been dramatic. The drive to connect additional nodes to the network is constantly increasing, while the amount of addressable space in the current protocol scheme continues to diminish. While it may seem that the 4.3 billion addresses provided in IPv4 (Internet Protocol version 4) is rather extensive, a large number of these addresses have already been delegated to large-scale users or reserved for highly specialized networking purposes. Therefore, the number of new nodes that can be connected to the Internet under the current addressing and networking scheme is shrinking rapidly [1].

On top of the address space issue, there is the increased amount of load placed upon routers and the network infrastructure required to process IPv4 packets. There is excessive complexity introduced by several unnecessary parameters, specifically error-detection codes. Security concerns, such as source address verification, have recently become major issues. These problems have been addressed by the development of a new Internet Protocol, known as IPv6 (Internet Protocol version 6) [2].

▲

In your literature review you should also include any material you find that would not support your ideas and discuss why you would like to proceed with the project despite this evidence. As an example, products have often been introduced before the technology has matured. If the products had inconvenient interfaces, were too expensive, or were too slow, consumers would not purchase them. Later, when the technology improved, the products were reintroduced with great success. If you plan to work on an existing idea that was unsuccessful in the past, then you should find literature explaining why the idea was unsuccessful and discuss how you think you can improve the product or design so that it will be successful now.

You should start your literature search with engineering society journals such as the proceedings and transactions of the Institute of Electrical and Electronics Engineers (IEEE) and reputable magazines and trade journals such as *Science, Nature, Analog Integrated Circuits and Signal Processing, Power Engineering,* and the *EE Times.* If your library does not have copies of these on hand, a reference librarian should be able to help you to obtain any necessary copies. An oft-repeated word of warning: if you find an electronic reference to a printed document, then you should have your library obtain a copy for you, so that you can check the reference. Unless the documents are official publications of a group such as a professional society or a standards body, there is no peer review on the Web, as there is with many printed documents. No one checks whether the references are correct, whether the methods used in obtaining data were standard, or whether the conclusions drawn were reasonable.

*Design Approach and Method of Solution*    This section is central to your document and must answer questions 4 to 13 as well as any questions that pertain particularly to your project. This is the section in which you explicitly list the design alternatives and discuss the details of your design. A summary of design alternatives can be found in the *Coreware* final report in the Appendix.

Often this section is divided into many subsections. For instance, if your solution has multiple components, such as hardware and software, or electrical and mechanical subsystems, each of which would require more than a paragraph to discuss, then you should break the solution into component subsections. An example of design subsections for an electrical engineering project can be found on pages 7 to 9 of the draft final report *The Talking Book,* which is included in the Appendix.

*Economic Analysis*    An economic analysis of your project answers the question, "Is the project feasible?" As discussed in Chapter 5, there are three types of economic analyses that can be performed, each with increasing amounts of detail. The one of immediate importance is the budget for the prototype, that is, what it will cost for you or your department to complete the project. For an undergraduate design project, this is your out-of-pocket expense plus any materials, equipment, or software provided by the university or an external source. An industrial project budget is slightly more complex. This is typically more extensive and expensive since your salary and computer or manufacturing time must be accounted for (computer time at universities is free and there are often free or lost-cost machine shops available, whereas in industry, particularly if you are not working for a small start-up, computers and equipment are dedicated to departments and must be leased from the departments). Alternatively, you could create a budget as if you were starting from scratch — an entrepreneur's budget that would include all costs associated with starting a company, purchasing equipment, paying salaries, and marketing the final product.

As shown in the examples in Chapter 5 and the reports in the Appendix, budgets are usually displayed best in the form of tables. However, it is insufficient to simply put a table in the report and assume you are finished. Each table must be accompanied by a discussion of the highlights. For instance, you should mention all of the assumptions made in the budget. You may choose to discuss the "big ticket" items, and consider how these costs may be reduced. When discussing the budget for the prototype, you should explicitly consider how this project will be funded. For an undergraduate project, if the budget is small, can all team members contribute enough to build the prototype? If the budget is large, has your faculty or industry advisor agreed to support you? Are there grants or awards you can apply for, and how will you complete the prototype if you do not receive them? For an industrial project, will your department be able to pick up all of the cost? Can the cost be shared between multiple departments? Are there any mechanisms at your company to fund innovative projects?

In addition to the budgets, you might include a pricing or price constraint analysis. Depending on your project, you should answer either of these two questions: How much will it cost to manufacture your product in quantity (if applicable) and what is a reasonable price to charge a customer? Or, what will it cost to implement and maintain your design? That is, are there any cost constraints to the design? In either case you must convince the audience that your project is economically feasible and that you are producing a useful and cost-effective design that customers can afford. One of the ways to estimate production costs or selling price is to extrapolate from current designs. A second and often effective method of determining selling price or cost constraints is to conduct an informal market survey and break down in a table the numbers of people who would be willing to purchase your product at given prices. (Market surveys are mentioned in Chapter 5.) If you perform a market survey, you should include a copy of the survey and details on the results in an appendix.

*Time Line or Gannt Chart*    Before you begin any project you must have an idea of how long the project will take. All documents must contain a time line or Gannt chart as discussed in Chapter 3. There are two options for presenting a chart. The entire chart (or series of charts) may be placed within the body of the proposal or report. If the document is severely page-limited, you may place it in an appendix and provide either a detailed discussion or an abbreviated version of the chart in the document body. Whether the time line is in the body of the document or the appendix, you must discuss it. Provide at least an overview of the major deadlines and the tasks requiring the longest times. One of the easiest methods of presenting a time line is to reproduce the major milestones in a small chart, discuss this in the body of the document, and then refer the reader to an appendix for the complete time line. The following examples contain reduced (Figure 6.4) and detailed (Figure 6.5) charts, respectively. These two examples are taken from the proposal of the *Coreware* project and show the planned activities for the three-term project that began in Fall 2000. Notice that the reduced chart did not explicitly include delivery dates or major milestones, which is a feature your charts should include.

If the document is an interim or final report, then you must discuss any time line changes since the last report. During the project, it is important to convince your audience that you can recover from any time line slips resulting from design modifications or challenges encountered. After a project is complete,

| ID | Task Name | Start | End | Q4 00 | | | Q1 01 | | | Q2 01 | |
|----|-----------|-------|-----|-------|----|-----|-------|-----|-----|-------|-----|
| | | | | Oct | Nov | Dec | Jan | Feb | Mar | Apr | May |
| 1 | Initial Preproposal | 10/11/2000 | 10/18/2000 | ▮ | | | | | | | |
| 2 | Final Preproposal | 10/18/2000 | 11/1/2000 | ▬ | | | | | | | |
| 3 | Architecture Design | 9/27/2000 | 11/14/2000 | ▬▬ | | | | | | | |
| 4 | Component Design and Testing | 11/1/2000 | 2/1/2001 | | ▬▬▬ | | | | | | |
| 5 | Proposal - Written and Oral | 11/14/2000 | 12/4/2000 | | ▬ | | | | | | |
| 6 | Component Integration | 1/1/2001 | 3/30/2001 | | | | ▬▬▬ | | | | |
| 7 | Progress Report | 2/14/2001 | 2/28/2001 | | | | | ▮ | | | |
| 8 | Integrated System Testing and Debugging | 2/1/2001 | 5/9/2001 | | | | | ▬▬▬▬ | | | |
| 9 | Final Report | 4/9/2001 | 5/9/2001 | | | | | | | ▬▬ | |

**FIGURE 6.4**
Reduced Gannt
chart.

you typically review the project to learn how you can improve your design strategy for the next project. Documentation of the time line changes is a good way to evaluate your performance.

*Societal and Environmental Impacts*   Whether or not you are required to explicitly produce a section on the societal and environmental impacts of your work, you should consider and discuss these impacts. Chapter 2 provides a list of questions to ask yourself. Consider each carefully. It is not necessary to provide an answer to each of the questions in your proposal or report; instead, consider those that are applicable and carefully provide answers. For instance, you may be producing a product that is biodegradable, has no moving parts, and is too large to be swallowed by a toddler. With a product such as this, it is easy to boast about the lack of environmental impact and the safety of the product, and you should definitely expend at least a few sentences on this. However, if the manufacturing of your product produces significant waste material, or if your product is likely to put many people out of work, you should discuss this as well. As discussed in Chapter 2 it may help to refer to sections in the code of ethics of your professional society.

*Summary, Conclusions, and Future Work*

After the abstract or introduction, this is most important piece of your document. Typically, if your reader has limited time, he or she will only read these two sections. Therefore, you should always include a "Summary" section, and this section cannot be an afterthought. This section gives you the opportunity to reiterate the high points or advantages of your project. You should summarize the major accomplishments you made and the benefits of your project. There is no need to go into great detail, since you have covered the design or proposed design already, but you should briefly cover the specifications, highlight the significant advantages of your work, summarize what you have done, and, particularly for reports for class assignments, summarize what you have learned.

If you were unable to finish a part of the design process, you should mention that fact here. If, upon reflection, you decide that one of your design alternatives would have been a better approach, discuss why and how modifications to the project could be made.

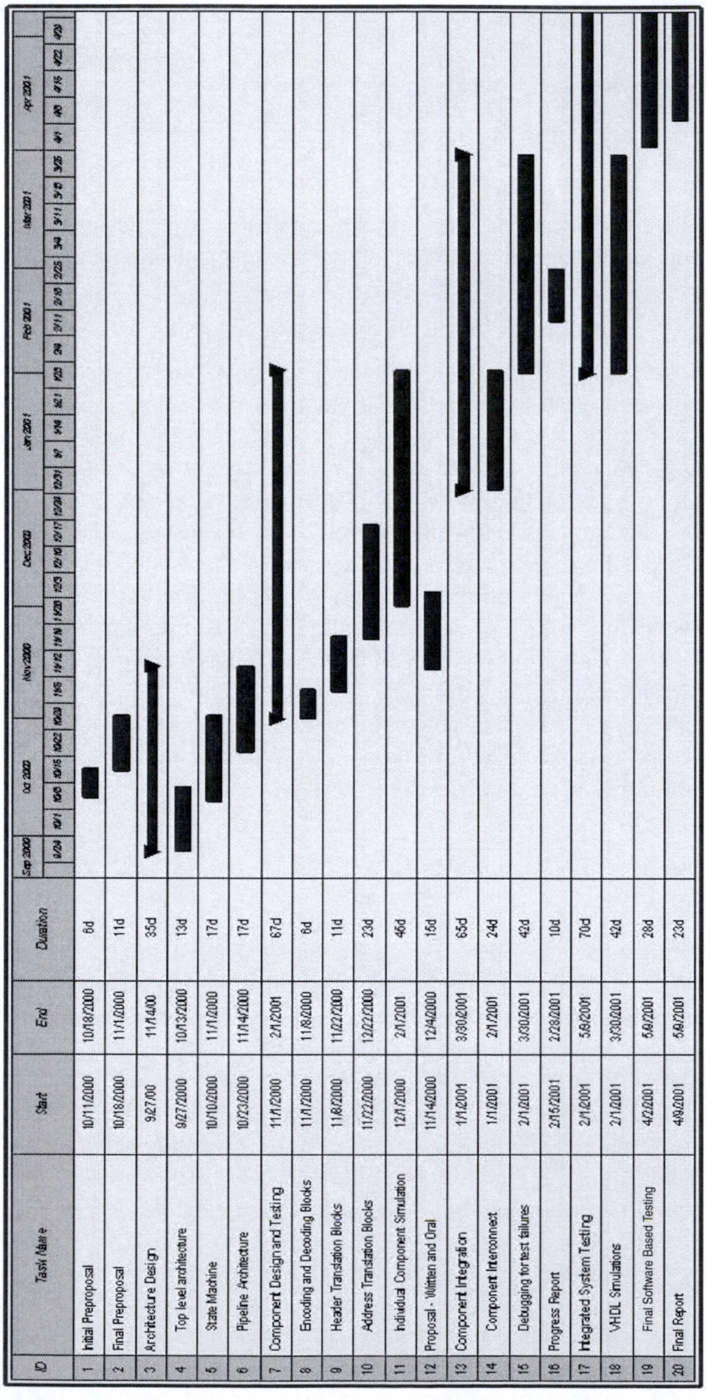

Practical
Engineering
Design

**FIGURE 6.5**
Expanded Gannt chart.

Finally, if permitted, you can include a "Future Work" section. This is your chance to let your imagination soar. Consider how you would extend your project if you had more time or money or better technology. Future teams will read your report and may continue with your work, using your project as a starting point.

## Appendices

Given the typical page limits of the body of a design report, you should expect to have multiple appendices. Any figures, tables, or graphs can be placed in appendices as long as they are discussed in the body of the report. Examples of what could be placed in appendices are:

- Glossary
- Gannt chart
- Detailed budgets
- Formal test procedures
- Test results
- Circuit diagrams
- Flowcharts
- In-depth explanations of technology or components being utilized
- Code
- Images of product or components
- Screen shots of software output
- Screen shots of the user interface
- Pictures of the prototype
- Line drawings

This list is long and by no means complete. Everything that is related to your project, that would help to explain your project or support your decisions, should be placed in an appendix and carefully labeled, with explanations.

Note the extensive appendices of the *Coreware* project in the Appendix.

There are several methods for delimiting appendices. The first is to separate each appendix by a title page giving the name of the appendix. This method is typically used when appendices are long, much longer than a page. Alternatively, you may choose to label each appendix at the top of the page as was done in the *Coreware* documents. This is typically done when appendices are very short, on the order of one page each.

## Proposals vs. Status Reports vs. Final Reports

The differences between the proposal, status report, and final report are the type of information presented and the organization of the information. The proposal focuses on the motivation behind the project, the patent and literature search, and the approach. (See the *Coreware* proposal in the Appendix for an example of proposal outline and tone.)

A status or progress report is used to update a manager or advisor about the status of the project, so there is less need to focus on motivation and background, but more need to focus on the Gannt chart and project tasks. However, it is useful to provide some background information on the project in order to refresh the memory of a busy manager. (An example status report style is given in the *Coreware* status report in the Appendix.) In this example, the team chose to provide background information on the project goals and the proposed and alternative approaches. The crux of the report is in a section titled "Progress Toward Solution," with a discussion of work remaining in a section titled "Future Work." The Gannt chart is placed in an appendix and discussed at the end of the report.

An outline of an alternative method of presenting report status is given below. In this method the focus is on the Gannt chart and the budget. The Gannt chart is placed up front and major milestones are immediately discussed. Each major task and important subtask is then discussed in turn, with mention made of the progress of each. In the case where there were unusual circumstances (obstacles overcome or alternative solutions employed), these should be mentioned. Note that obstacles overcome are not problems such as lack of knowledge of an operating system or a tool, since you might reasonably be expected to know these or be able to learn this easily.

▼

### SAMPLE OUTLINE OF STATUS REPORT

| Section | Approximate Page Count |
|---|---|
| Abstract | 1 |
| List of Tables | 1 |
| List of Figures | 1 |
| List of Abbreviations | 1 |
| Introduction | 2 |
| Gannt Chart | 2 |
| Overview of Major Milestones | 1 |
| Discussion of Task Progress | |
| Task 1 | 1 |
| Subtask 1.1 | 0.5 |
| Subtask 1.2 | 0.5 |
| Subtask 1.3 | 0.5 |
| Task 2 | 0.5 |
| Task 3 | 1 |
| Subtask 3.1 | 1 |
| Subtask 3.2 | 0.5 |
| Task 4 | 1 |
| Updated Budget | 2 |
| Discussion of Budget Modifications | 2 |
| Appendices | 5 |

▲

For the final report you should again repeat the format of the status report, in the sense that you must give an overview of the project, the motivation, and outcomes. It is no longer necessary to discuss the Gannt chart in depth; however, you might include it in an appendix. It is necessary to discuss in detail the operation of the hardware or software, the test outcomes, and also to present end-user documentation. You can also discuss design alternatives that may be promising.

### 6.3.5   Test Plan

Designers often tend to leave testing until the end of a project cycle. This is poor practice, since finding incompatible components or finding that users cannot understand your interface could ruin a project. A well-organized, successful project has an integrated testing that follows a detailed test plan.

A test plan is a detailed plan describing the components or interfaces to be tested, the test procedure (steps in the test), and the expected results or range of results. As described in Chapter 1, all components should be tested independently, and then systems and subsystems should be tested again after integration. Finally, the entire product should be tested for performance as well as for usability. At minimum, the test plan lists the test steps, the order in which the tests should be performed, and the resources (people, equipment, and software) needed for the testing. For industrial projects, there is often a specified format and list of test metrics and counts.

---

### SAMPLE OUTLINE FOR A MINIMAL TEST PLAN

Project Overview
Test Schedule
Test Materials and Equipment
Test Procedure Overview
    Component Testing
    Integration Testing
    User Testing
Components
Integration Testing
User Testing

---

See Keyes (2003) for an overview of software testing and a sample of a detailed test plan.

The specifics of the plan will depend on the type of product you are designing. As an example of a hardware project, let us again consider the Widget, a consumer product that we will assume is to be used by people of all ages. The following example shows an initial test plan that you might develop as you are evaluating your topic for feasibility. It is little more than a checklist that will remind you of what will have to be tested as you are in the process of designing the Widget and can easily be expanded upon as you proceed through the project. Note that this is by no means comprehensive; the "Operation" category should be expanded upon to include testing the interoperability of each of the components in the design. For projects other than electronic consumer products, you would have to modify this initial plan accordingly.

---

### SAMPLE TEST PROCEDURE OVERVIEW FOR THE WIDGET

The Widget will be tested to ensure that it meets the design requirements in each of the following categories:

Operation
    All developed or purchased components individually meet design
        specifications.
        <List components and specifications.>
        <Refer the reader to the individual component test plan and test
            results.>
    The product operates according to specifications.
        <List specifications.>
        <Refer the reader to the individual component test plan and test
            results.>

Safety

Casing is durable with no protruding edges or components.

<Refer reader to tests on breakage pressure.>

There is no risk of electrocution or fire hazard if used improperly.

<Refer reader to tests current at all pertinent locations throughout device.>

It is sufficiently large so that there is no choking hazard.

<Specify size and refer to literature on hazards.>

Durability and Efficiency

All components are securely mounted.

<Refer reader to tests on breakage pressure.>

Power consumption is sufficiently low so that battery life meets design specification.

<List power consumption specification.>

<Refer reader to the power consumption test plan and test results.>

The product lifespan meets the design specifications.

<List specifications.>

<Refer the reader to the individual component test plan and test results.>

▲

The example below shows a detailed test plan for determining the power consumption of the Widget. Note that this procedure is obvious, but it will again serve as a checklist to help ensure that you have adequately tested the product. If the Widget had two modes of operation, one that requires more power than another (for instance, think of the difference between the battery life of a laptop that is on standby mode vs. one that is constantly performing writes to the disk), then each of these modes should be tested separately. The power consumption should then be listed for each of the modes and the predicted average power consumption calculated.

▼

## SAMPLE DETAILED TEST PROCEDURE FOR WIDGET POWER CONSUMPTION

Description:

The purpose of the test is to determine the nominal operating time of the Widget using the specified batteries. Expected duration of operation is 32 hours and is critical to the success of the project.

Test Steps:

1. Insert fully charged batteries.
2. Turn Widget on.
3. Check every 30 minutes whether Widget is operating.
4. Record battery duration.

Expected Result:

32 hours or greater duration

▲

▼

### PRELIMINARY DETAILED TEST PROCEDURE
### FOR A SOFTWARE MULTIPLICATION ROUTINE *NEWMULT(X,Y)*

Description:

The purpose of the test is to determine whether the routine can accept two integers and output the product of the integers. The test will focus on ensuring that the routine returns the correct number for a specified range of integers and successfully identifies integers that cannot be multiplied.

Test Steps:

1. Define x and y as integers.
2. Set x = 1, y = 2.
3. Call routine newmult(x,y).
4. Record output.
5. Repeat steps 1–3 for integers << maximum of system.
6. Set x equal to maximum integer allowed by system, set y = 1.
7. Call routine newmult(x,y).
8. Record output.
9. Set y equal to maximum integer allowed by system, set x = 1.
10. Call routine newmult(x,y).
11. Record output.
12. Set x equal to maximum integer allowed by system, set y = 10.
13. Call routine newmult(x,y).
14. Record output.
15. Set y equal to maximum integer allowed by system, set x = 10.
16. Call routine newmult(x,y).
17. Record output.
18. Set y and x such that their product is greater than the maximum integer allowed by the system.
19. Call routine newmult(x,y).
20. Record output.

Expected Result:

For integers less than the maximum value allowed by the system and whose products are less than the maximum allowed by the system, the routine outputs the correct value of the product in integer form. For integers greater than the maximum value allowed or whose product is greater than the maximum allowed, the routine outputs an error message.

▲

These examples illustrate simple tests; however, even integration or user testing can be broken down into straightforward procedures such as these. Note, in particular, that the second test overview is useful in outlining tests you might forget if they are not written down; that is, it would be easy to select two small, positive integers and check to see if the routine outputs the correct answer. However, we see that we have to test the error-catching ability of the routine by testing with large numbers, numbers that exceed the system capabilities. By reading through this test and looking for parallels, we also observe that the negative integers were neglected. We should explicitly check negative numbers and the minimum (most negative) numbers the system can handle.

### 6.3.6  Test Report

The test documentation is not complete without a report (or at least a section) that describes the test results. The test report is written to inform the customer what tests were performed and whether all the expected outcomes were seen, or whether there were any unusual outcomes. The report should begin with an introduction describing the project and an overview describing the outcomes of the tests. For successful tests, the overview may simply state, for example, that beta testing (testing by external users) of the user interface was performed and the prototype passed all tests. If the prototype failed some of the tests, this must be pointed out. Likewise, if there were any unexpected results, for example, you expected the prototype to have a battery life of 36 hours in standby mode when the test showed that the life was 45 hours, these should be mentioned. Finally, if the system under test fails any tests and you are able to fix the system, you must report on the initial failure, the method of fixing, and the results of the new test that determined the success of the system.

The following example contains an outline of the minimum required for a test report. Note that most of these sections are already in your test plan; they can be copied and slightly modified for your test report.

▼

#### SAMPLE OUTLINE FOR A MINIMAL TEST REPORT

Project Overview
Summary of Test Outcomes
Test Materials and Equipment

Test Procedure Overview
    Component Testing
    Integration Testing
    User Testing

Components
    Component Test 1
        Description
        Expected Result
        Actual Result
    Component Test 2
        Description
        Expected Result
        Actual Result

.
.
.
.

Integration Testing
.
.
.
.

User Testing
.
.
.

▲

## 6.3.7 User Documentation

All projects need user documentation, if only to describe the number and type of batteries needed and the operation of the "on" switch. However, almost all products will require more documentation. The user manual may range from a one-page pamphlet with helpful graphics describing the operation to documents the length of long novels describing the inputs and output limits and formats of each software subroutine along with instructions for installing the software. In the past you have most likely encountered user documents that were unreadable (perhaps because they were translated poorly or perhaps because the documents were simply poorly written). You may recall the frustration of trying to get what should be a simple operation to work, whether assembling a product or installing a software package. From this experience you already know that it is vital to your project to have understandable and correct instructions that cover all usage scenarios. You cannot have satisfied customers if they cannot use your product.

While user documentation will certainly depend on your product, a general outline of a document is given below.

▼

### OUTLINE OF A SAMPLE USER DOCUMENT

Introduction and Overview of Product
Intended Users
System Requirements and Specifications
Safety Issues
Detailed Installation or Assembly Instructions
Operation
Contact Information or Support Services

▲

The introduction is a paragraph to multipage overview of the product, listing the purpose of the product and the uses of the product. "Intended Users" is a vital section. You must answer whether there is an age range for your product. For example, is your product suitable for children ages 3 to 6 or is designed to assist people with certain disabilities? Also, if your product calls for specialized knowledge you must state that here. Is the product designed for users with particular certifications, such as machine operation or software certifications? The system requirements are suitable for both hardware and software products. For hardware products, a specification might be a particular input voltage and current. For software products the requirements might be a specific platform, that is, specific computer hardware (2 GHz processor, specific sound card) or a specific operating system (Linux, Windows CE, or DOS).

Safety issues are particularly important. These issues range from "Do not operate the equipment near water" or "Equipment must be unplugged before changing blade" to computer security issues: "If you have platform X you must install using this option or your system may be vulnerable to data loss."

The detailed assembly and installation instructions are equally important. The operation section would be broken down into subsections specific to your project. For instance, *The Talking Book* would have instructions for inserting batteries, the procedure for recording reading, and the procedure for playing the recording. Software would have instructions for using the graphical user interface (GUI) or lists and details of command line instructions, and for each routine that a user can access, a description of the routine's function, a complete list of valid inputs and outputs, and a list of error messages.

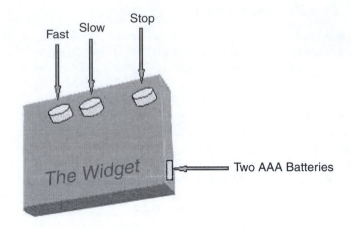

**FIGURE 6.6**
Simple product illustration.

Good user documents typically contain graphics that assist with explanation. For instance, you could use a screen shot of a GUI window to illustrate where the various pull-down menus or tabs are located, or you could have a line drawing of your product to indicate where the battery access door is and the locations of the switches. Figure 6.6 shows a simple line drawing of a product that clearly shows the product orientation and which buttons perform which operations.

Testing your user documentation is essential to ensuring that you have included all necessary steps and directions. While documentation testing should be performed separately from product testing, it is possible to combine these two operations if you are strapped for time or short on people. In either case, you should develop a formal test plan for testing the usability and also ask your test subjects to serve as reviewers for your user manual.

## REFERENCES

Alred, G.J., Brusaw, C.T., and Olin, W.E. (2003). *Handbook of Technical Writing*, 7th ed., New York: St. Martin's Press.

This book is arranged like an encyclopedia with easy-to-find topics. It covers subjects from basic grammar to the layout of different types of documents.

Keyes, J. (2003). *Software Engineering Handbook*, Boca Raton, FL: Auerbach.

This book contains a chapter on both component-level testing and integrated testing that is applicable to both software and hardware designers. It also contains a detailed test plan that can easily be adapted to any project.

Pfeiffer, W. (2000). *Technical Writing: A Practical Approach*, Upper Saddle River, NJ: Prentice Hall.

A solid general reference containing a good chapter with many examples of using and creating graphics.

## FURTHER READING

Markel, M. (1994). *Writing in the Technical Fields: A Step-by-Step Guide for Engineers, Scientists and Technicians*, Piscataway, NJ: IEEE Press.

An excellent general reference for technical writing; it is particularly useful as a design project reference.

Williams, J.M. (2003). *Style: Ten Lessons in Clarity and Grace*, 7th ed., Upper Saddle River, NJ: Pearson.

For those who are interested in further developing their writing skills, this is a short book with excellent descriptions of how to create different tones in writing.

★ *Key Points*
*Chapter 7*

- Presenting is much more than speaking to an audience.
- Planning is almost everything.
- Practicing is the key to confidence in your plan.
- Knowing your audience is as important as knowing your subject.
- Demonstrating teamwork is as important during the presentation as before it.
- Using good graphics is a given.
- Using glitz (particularly PowerPoint glitz) to disguise lack of substance can antagonize an audience.
- Do not read the slides, just elaborate on the bullet points.
- Having a back up plan is mandatory. Expect disaster to avoid it.
- Visiting the presentation room in advance is helpful.
- Knowing the answer to every question is unlikely; knowing how to answer is crucial.
- Showing enthusiasm gives the audience confidence in you.
- Dress for success; you have earned it.

# Chapter 7

# PRESENTING YOUR DESIGN PROJECT

*Valarie Meliotes Arms*

## 7.1   INTRODUCTION

Making oral presentations effectively requires planning from the beginning of the project, practicing with an audience, and controlling the actual presentation. Taking each step seriously is a good investment in your career. Presenting your design project prepares you for a crucial job skill. Industry has actively encouraged universities to promote oral communication skills because of the perceived weakness of engineering graduates in communicating within their own field and beyond — to co-workers, managers, and potential clients.

The largest aerospace company in the world, Boeing Corporation, has invested in a program with the objective "to influence the content of engineering education in ways that will better prepare tomorrow's graduates to practice engineering in a world-class industrial environment." Accordingly, Phil Conduit, the CEO of Boeing, has identified the "4 Cs" engineers need to embrace: (a) collaboration, (b) communication, (c) cost awareness, and (d) continuous learning. The design project offers opportunities for collaboration and communication that will set you on the path to a successful career as an engineer.

## 7.2   PLANNING

### 7.2.1   Who Is the Audience?

Planning does not start when you realize the design presentations are next week. Planning starts when you start working on the project. From the moment you define the problem, you should be defining a problem to a specific audience. The audience has to be a real entity in your thinking because engineering is in many ways a social act. You may have a naïve audience who does not understand the technology involved, forcing you to explain in simple language. Engineers have shaped the world with technology but they do not manufacture, market, or exclusively use what they design. The public that does is the audience that funds the engineering enterprise.

Thinking of the audience's responses to your design may be even more crucial if you work in a company that makes oral presentations rather than submitting written proposals. Development companies which must address constituents from politics, the community, and venture capital often make oral proposals first to generate the questions and concerns that must be addressed in a written document.

In a design presentation, you may be faced with an audience that includes your advisor, managers, venture capitalists, or company shareholders. You must be able to speak to each member of your audience with persuasion. It is wise to practice with a real audience so enlist colleagues, family members, friends, and of course, your advisor or supervisor, for rehearsals. Do not dismiss any questions because they will tell you how you are reaching an audience like the one you will encounter at the presentation. You may discover that a silly question indicates a vague area that could be better explained. If it really is silly, it allows you to gather experience in responding diplomatically.

### 7.2.2 What Should the Audience Do When You Finish Presenting?

Once you have considered the makeup of the audience, you should consider what you want them to do as a result of hearing your presentation. The answer may be as simple as "Give us an A!" or "Give me a good review" or as complex as "Help us enlist government aid or industry funding for an environmentally touchy project." The answer you want will determine the order in which you present information. For a coursework project you should ask to see the evaluation form by which you will be judged to decide what to emphasize. The department probably has a grading sheet that indicates areas for points to achieve 100%. Company evaluation forms, like those of major funding organizations, should tell you what to cover, so do not overlook the obvious, no matter how engrossed you are in the technical details.

If you need to persuade an audience to take action, you should order the information in a way that will motivate them, and that clearly depends on who they are, not on who you are. Once you convince them of their need for your project, then you can discuss the technology to achieve a solution and your team's expertise to do so. That means you go back to the chapter on design (Chapter 1) and consider the need for a solution to a carefully defined problem, the alternatives, the well-documented research that leads you to a proposed solution, the criteria by which to evaluate success, the societal and environmental considerations, and the cost. The emphasis of a particular presentation is determined by the audience you are trying to persuade.

## 7.3 CREATING A PRESENTATION

### 7.3.1 Slides

Assuming you have considered how best to present real content to real people, software such as PowerPoint can guide you through creating key words with bullets to indicate points worth remembering. While PowerPoint has its dangers, it offers the advantage of helping to overcoming inertia. Having to think of key words for headings and bullets for details can get you going in the right direction.

However, the glitz offered at your fingertips with AutoContent, animation without purpose, and busy backgrounds poses several hazards. In his article "Absolute PowerPoint" (Parker, 2001) Ian Parker writes, "It's easy to avoid these extreme templates —many people do." Unfortunately, many people do not. Sound effects that are inconsequential, color that is distracting, clip art that is pointless — all have become common substitutes for content. The result is likely to be an irritated audience. One of the developers of Microsoft Office illustrates this point with the story of what happened when a speaker at a conference had a projector fail: "Everybody was ... cheering. They just wanted to hear this woman speak, and they wanted it to be from her heart. And PowerPoint almost alienated her audience." Certainly, using PowerPoint correctly can be powerful. It also indicates that you have spent time finding appropriate images and key words. So think through the steps in planning. Choose images that show the audience what you want them to understand and select key words that truly highlight ideas, which you verbally explain.

The audience wants to feel that the emphasis is on their needs. Can each and everyone see the screen, hear the speakers, and feel their concerns have been addressed? To answer the audience's needs, you need to do more than put three bullets under a topic heading, as the usual slide framework suggests. You should know your audience well enough to make analogies to what they already understand. Analogies build bridges between topics and are safer than jokes to draw an audience to the speaker. While you should generally avoid jokes, you should seek out analogies that relate something you are talking about to something the audience already knows. Mortimer Adler, who wrote *How to Read a Book* (Adler, 1940), quickly captured attention with the analogy that reading a book is like eating a steak; you may have the steak in your refrigerator, but until you cook it, chew it, and digest it, it is not part of you. That analogy is unexpected, but adds interest and clearly relates the new idea to the known concept. You can create analogies with a little effort. If you remember a poem that captured your attention with a striking image, think about your project that way. It is a proven characteristic of creative thinkers to be able to make analogies between different or unlikely fields. It requires a willingness to play with ideas that will be far more impressive to an audience than a lame joke they may have already heard or recognize as filler for want of creative thinking.

As an example of addressing the audience's needs, in Figure 7.1, a professor has presented material that illustrates several points. He has used humor pertaining to his topic, scheduling adequately for senior design. He has several items on the list and he intends for the audience to read these on their own. The type is sufficiently large to allow for the author's preference for fonts with serif. If you are sure the slide is visible, then it is acceptable. The Gantt chart is not meant to be read in detail but rather to lead the audience and the lecturer to the next topic.

### 7.3.2 Backup

Whatever you select as the primary mode for visual presentation of the project, have at least one form of backup. In most instances today, the primary mode is a computer slide show that has many built-in aids to provide a framework any novice can use. But computers are even more prone to glitches, from humans and gremlins, than overhead projectors whose bulbs burn out. If you use a computer slide show, print a set of overhead transparencies for emergency

**FIGURE 7.1**
From a lecture on
scheduling. (From
Prof. J. Mitchell, Lecture
on Project Management.)

### How Can You Avoid These Problems

❖ Accept whatever grade you get and have a good time
❖ Spend all your time on Senior Design
❖ $1,000 (negotiable) in small, unmarked bills
❖ Plan!

use. If the computer crashes and you have to use the old overhead projector, at least you will impress the audience with your ability to deal with a crisis, a desirable ability in any field of engineering. If the computer is your laptop with the slide presentation, have a separate CD or diskette with the presentation that might work on another computer, should one be available. Also carry the cables necessary to connect your computer to both the projector and the power supply. If all this seems extreme, ask anyone who has given presentations to describe the problems they have witnessed.

If possible, attend a few presentations in advance of your own to witness what can and does go wrong. Cables disappear, systems crash, someone gets stuck in a traffic jam (guess who has the graphics?), prototypes mysteriously implode, and advisors demand, "Where's your backup?" Obviously, if you are traveling to do a presentation, you must have not only a backup of the slides but multiple carriers. That includes having more than one copy of your presentation among the team; accidents can and do happen, whether you are traveling across campus or across country. In industry the consequences can be dire. An engineer on his way to another city to pitch a multimillion dollar proposal watched his partner board a plane that he was too late for and he could do nothing to get the slides for their presentation into his partner's hands. Because they had rehearsed together, the engineer who made the day trip was able to cover the missing material, but he admitted they did not get the job for which they were bidding. To ensure your success, make your policy firm: one person carries the CD with the PowerPoint presentation, a second carries the laptop that has the same presentation, and a third carries the transparencies. Everyone arrives with fewer butterflies in their stomachs.

### 7.3.3 Format

Many presentation programs suggest the font and size automatically, and these are only worth varying when you are absolutely confident of the material. The goal here is to impress the audience with your efforts in design, not in graphic artistry. Generally, the screen should have three or four major points in a type that can be read from across the room you will be using. As in your written work, use specific words rather than general ones to produce as much impact as possible in a little space. "Improve productivity" is generic; "Find chip costing less than $30" is more specific, therefore more memorable. Everyone has an "Introduction," "Discussion," and "Conclusion," if you want to state the obvious. With a little planning, you should find headings that describe your project and none other.

Always check the spelling and the grammar and have someone unfamiliar with the presentation check them too. You may become so accustomed to the ideas that you stop reading what is on the page and substitute what is in your memory, never noticing any discrepancy.

▼

### CHECKLIST FOR FORMATTING

- Check calculations for common sense results.
- Be consistent with numbers throughout the report.
- Check grammar and spelling; they still count.
- Put units for all numbers (Is that a foot or a meter?).
- Avoid black type on a deep-blue background; it is illegible.
- Define acronyms and jargon if you must use them.
- Use numbers only to significant digits.

▲

Font style and size make a difference. For best readability on any type of overhead, use sans serif fonts meaning "without the feet and the tail." The extra space or whiteness gives a less cluttered look than serif font styles, which, like cursive writing, seem to flow — appropriate in hard copy but not for bullets or talking points in your presentation. Likewise, to ensure readability you must not use similar colors such as black with dark blue or yellow with white. In general, avoid shades of the same color on the same slide. In all cases, check to see that the colors are visible from the back of the room.

The advantage to bullets of information is that you avoid reading lines of text. It is insulting to an audience to have the speaker read what they can plainly see for themselves. If you have bullets, you can elaborate on them; if you have a diagram you can use a pointer to indicate key areas. Be sure you know what is on the screen and face the audience to explain it. If feasible, stand to the left of the screen from the audience's point of view, as the tendency to read from left to right will lead their eyes back to you and to the starting point of the next bullet. Glance at each screen quickly as it appears to guarantee it is the right one and to remind yourself of what to say next, then regard your audience as you expand on it. If you have practiced, you will avoid the common error of turning away from the audience to read what you should know by heart after months of work.

If you use data or images from sources other than your team, they must be acknowledged. Be sure to print the necessary information under the graphic. Pictures, like words, are intellectual property that must be recognized. It is also smart to acknowledge sources since they give your data the stamp of authority. For example, if the Environmental Protection Agency or the Institute for Electrical and Electronics Engineers has conducted research or collected data, their status gives a ring of truth that you as a novice researcher may lack.

### 7.3.4 Parallelism

Bullets are inherently prone to the poor parallelism syndrome. Some readers may never notice, but this one little grammatical fault invites charges from sloppy thinking to sloppy formatting. When clarity is critical for your audience to understand and remember key points, poor parallelism can be confusing and misleading. Parallelism is important in writing, but even more so in speaking because it provides a mental clue to how to organize the information.

Parallelism is simply using like parts of speech to start each item in a list. If a list begins with a positive action, as the "Checklist for Formatting" above does, then each item must follow that pattern. The first word must be a verb and it must command a positive action. The audience will understand and remember the list better than if some items switched to negative. A planning draft might have read, "Do not use black type on a blue background." While the point is the same as the bullet in the checklist, it does not fit the pattern and could cause confusion if the reader carries over the negation to the next bullet, the positive command "define."

Slides with bullets must be structured so that each element carries equal weight. If the first bullet is a noun, then each successive bullet must be a noun; if a verb, then the verb must be followed by verbs, etc. In the *Coreware* project, the students neglected to consider parallelism. The slide in Figure 7.2 looks fine but its authors may have been mislead in their thinking by trying to maintain a large header and then smaller bullets without regard for a subset.

*Parses*, *outputs*, and *determines* are strong verbs for three of the bullets. However, it is not clear whether these verbs describe the first bullet, "32-bit Information Bus," which is a subset of the title, "Decoder Blocks." If they do, then that point is lost, and the slide should indicate that the verbs represent properties of the Bus. If this is not so, then the first bullet should be changed to indicate that these bullets are all properties of the decoder. A similar failure to construct a parallel set of bullets occurs in Figure 7.3.

Again, strong verbs begin each of three points that represent properties. The first bullet is an explanation of the header. The structure would make more sense if the heading read, "Encoding blocks: Perform the inverse operation of the decoder." Other slides in the *Coreware* project have similar lapses in parallelism. The clarification would make the presentation more accessible to an audience struggling to understand complex flowcharts such as the one in Figure 7.4, which uses these two concepts.

In a real sense, the audience member who noted the problems with parallelism detected a lack of attention to the audience in general so that by the time the slide "Pipeline Timing Diagram" appeared, any desire to understand its complexity had eroded. That slide may have been necessary, but it requires notes, color, and perhaps a buildup of information to be layered as the explanation progresses. If you want to impress the audience, you must consider the effect of good grammar and good graphics. It is likely that good engineering is being presented by each contender for a proposal bid; it is unlikely that each will pay as much attention to the quality of their presentation as they do to the engineering.

## 7.3.5   Resources

You may have access to a graphics facility or commercial copy center that can help you in developing ideas or using equipment for the visuals. These places can get very busy because everyone waits until the last minute to call upon them. If you have been following the suggestions for scheduling you have planned ahead to avail yourself of their help well before the hordes cause long lines at crunch time. You will be calmer and better prepared if your Gantt chart includes a visit to such a resource a few weeks before the presentation. Ask about their resources, the cost, and turnaround time. If you are not ready to have the visuals produced you may at least be able to reserve time for their production closer to the presentation. People appreciate being able to plan their workload, and they may be able to accommodate last minute changes if you are already a client.

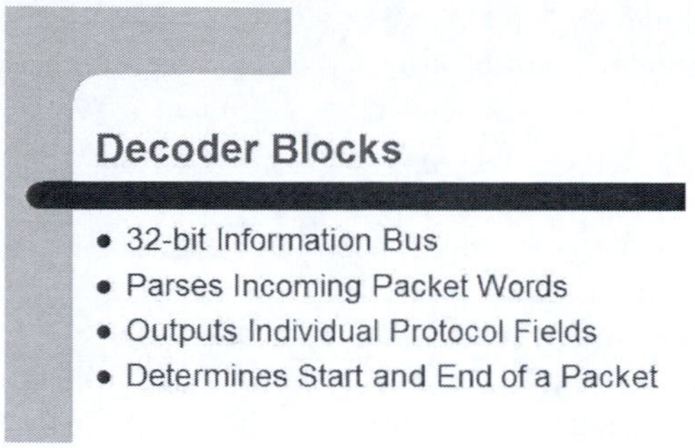

## Decoder Blocks

- 32-bit Information Bus
- Parses Incoming Packet Words
- Outputs Individual Protocol Fields
- Determines Start and End of a Packet

**FIGURE 7.2**
Parallelism example.

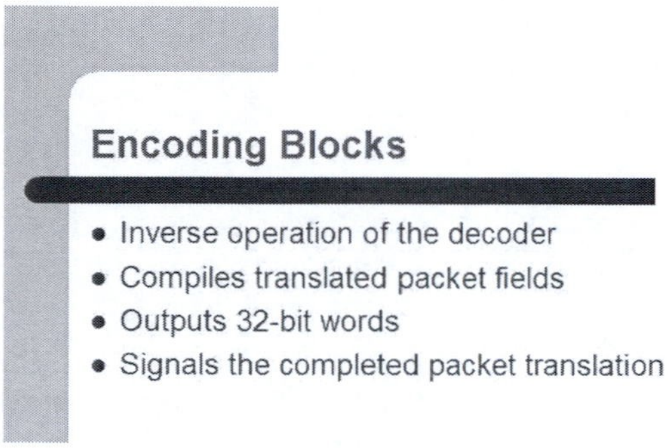

## Encoding Blocks

- Inverse operation of the decoder
- Compiles translated packet fields
- Outputs 32-bit words
- Signals the completed packet translation

**FIGURE 7.3**
Parallelism example.

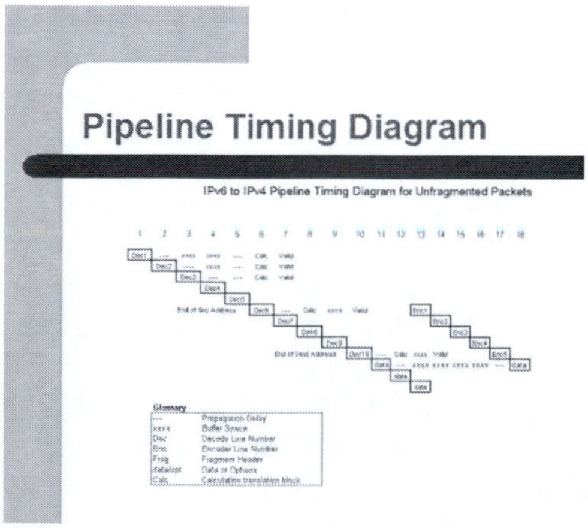

## Pipeline Timing Diagram

IPv6 to IPv4 Pipeline Timing Diagram for Unfragmented Packets

**FIGURE 7.4**
Pipeline timing diagram. A
complex diagram requires
clear explanation.

Your proposal and report may have graphics such as diagrams, flow charts, and spreadsheets. Resist the tendency to pluck these from the written document and simply project them. While they are useful to a reader with time to scrutinize them and refer to the text for your discussion, they are probably too busy for a listener 25 feet away from the screen and are typically in an inappropriate format (usually much smaller) than what is needed for a slide. You may find that you can simplify them and then build them in layers to achieve the complexity they have in a written document. They would become a good resource if you deconstruct them or at least consider how much time the listeners need to gather all the information they contain. Consider using Table 2 in the draft final report of *The Talking Book* (see the Appendix) in a presentation. Even the title of the table is too much to be spoken easily. The point the authors wish to make is that "the trend for the book sales industry is increasing." A simple arrow diagram climbing to the right will make the point immediately. Someone who wants the detail in the table could refer to it. Remember that the actual written proposal or final report with all its details and data will be accessible to the audience. A printout of your slides is not necessary as a "take-away," although it may help jog a listener's memory. The slides, in whatever form, lack the detail of your vocal or written explanation and that is as it should be. Otherwise, why should the audience listen to you speak?

The key to determining what elements to use is comprehensibility and visibility from the other side of the room. No matter how small the room in which you are working, get up and walk to the other side of that room to look at your graphic. Many speakers have never seen their own presentation except "up close and personal." Your audience will be several feet away and formal. Do not judge readability of slides from the computer screen or the projection screen beside which you are standing.

## 7.4   PRACTICING

Practice is vital for producing a successful presentation. You should therefore practice as many times and with as many audiences as possible.

### 7.4.1   With the Team

Practicing affords you the opportunity to gauge an audience's reaction, to time the whole team's presentation, and to become comfortable with speaking your part. You must practice to know how much time it takes you to explain your project and also to coordinate the graphics, whether you are handling them yourself or preferably for each other. The speaker should be focused on the audience rather than the computer or overhead projector; obviously, helping each other with the graphics tells the audience that you have rehearsed and demonstrates teamwork.

### 7.4.2   By Yourself

To a certain extent, you must see the presentation as a play and consider your costume and staging. It will be "all for nought" unless you have something worth saying, but if you fail to consider the context in which you say it, the audience may not hear you. It is helpful to use a tape recorder or a camcorder so that you can hear yourself speaking and see yourself as the audience will see you. You will quickly catch the "aah's," "you know's" and body movements that should be eliminated. Becoming conscious of them is the key to eliminating them. You should listen to yourself to hear if your pronunciation is distinct. If you think it could be better, a voice coach can help you improve. Most students can improve simply by practicing with clarity as a goal. Likewise, distracting body movements such as slouching or jamming your hands in your pockets can be corrected if you have time to pay attention to them as well as to the content of the design project.

### 7.4.3   In Costume

Novice engineers may be unused to wearing the formal attire that a presentation requires. Practice in the clothes in which you will present because you must feel comfortable with yourself and your ideas to convey confidence. If you have any doubts about the appropriate attire, do not hesitate to ask. What passes for appropriate on the West Coast may not be considered appropriate on the East Coast. Often, competing ideas are so close in merit that the audience's confidence in the speaker is the deciding factor in awarding a contract. Yes, what you wear and how comfortable you are in it does matter. The standards on dress codes change from place to place. In one instance, a group of professional consultants from Co. X forgot the importance of pleasing the audience in all respects; they followed their company policy of presenting in formal business attire when the client had specifically said their policy was *casual,* not even *business casual.* The client listened to the experts from several consulting companies and awarded the contract to a group who they thought was a good fit. Co. X was deemed incompatible even though they had the expertise for the job. They had not listened to the client's simple request about attire and thus made their ability to listen to technical matters questionable.

### 7.4.4 With an Audience

You can often gather an audience by promising to review for another team and by asking your advisor or supervisor. Such an audience will respect your efforts to perfect your communication skills as much as your technical knowledge. Honest listeners will tell you if you are audible or if you are blocking the screen. Better yet, they will help you anticipate questions from the real audience. Your presentation cannot possibly cover everything in a brief time, though you must cover crucial information. Inevitably, you will have to omit details, and judging if you have omitted the right ones can be demonstrated with a rehearsal audience.

### 7.4.5 With Aids

Your audience has some interest in the topic, but you have become the expert. You are going to explain the design project that you have been working on for months in a very brief period. You will need to condense and simplify more than you think possible. Graphics, from tables to figures to pictures, can order an array of information. Complex engineering designs require visual representation. The real thing is even better than a graphic, but if "the thing" is too small, too large, too complex, or not yet built, then you need a graphical representation.

Your body's appearance and placement can affect your listeners' reactions. If the audience cannot see the great graphics, the effort in creating them is wasted. Practicing with the graphics means having someone check on their visibility in the room. If you cannot practice in the actual room, then find one the same size. Have teammates sit in various places to check for readability of the graphics. You should never have to apologize for type that is too small to be read in the back of the room. Also, be sure that your body is not blocking someone's view of the graphics. Use a pointer and stand near the screen to one side for the best visibility.

Never give the audience a handout or pass something around before you have concluded the presentation. Their heads will automatically drop to look at what is in their hands and you will have lost your command of their attention.

## 7.5 PRESENTING

Remember, this is a performance and you must remain in command from the moment you begin until you end. Any audience will find enthusiasm and confidence infectious. Rehearse with the team, rehearse by yourself, rehearse with the graphics. Then, if you do feel nervous, you will be so well prepared that you will function on automatic pilot.

Timing is important, and practice sessions with the group should have addressed the pace for the allotted time. A rule of thumb is that it should take, on the average, two minutes to go through each slide. Everyone on the team should have a speaking part. If one speaker is weak, then place that person in the middle of the presentation with good graphics. Team members should always look at the speaker and appear as interested as you want the audience to be. You should time each other and agree on a signal, perhaps one minute to finish, by tapping on a watch or whatever works for you without distracting the audience. Likewise, you should help each other with the graphics. But avoid saying "Next slide, please," which implies that you have not practiced together and that you are unfamiliar with each other's work. A simple nod of the head or wave of a hand can alert the person running the slide show to move on.

### 7.5.1 Beginning

When your group takes the stage, you do so literally: you greet the audience, introduce yourselves and your topic, and command the attention of everyone in the room for your allotted time. By virtue of being in front of the room, you assume that command, and most audiences will want to cooperate with the conventions of the theater. They will be quiet when you are speaking, they will pay attention as long as you speak audibly and clearly, and they will wait until you finish presenting to ask questions if that is the stated format. They will also clap politely when you have finished. Nonetheless, most speakers, including professionals, get butterflies in their stomachs at the thought of having to address a group formally. Being nervous is a good thing. It means you are smart enough to know what you are doing is important, either for a grade or a company's bottom line. If you have practiced, the audience will see your flushed face as a sign of enthusiasm, even if what you feel is terror. Literally, stop and take a deep breath before you begin to speak. Pause during your speech to breathe, as this will prevent you from speaking too quickly. Your listeners will remain in place.

When you begin a presentation, the first speaker should introduce the team. You may decide to repeat the names as you move from speaker to speaker. For the last slide, it is polite to acknowledge anyone who helped you significantly with the project. Then the last speaker may finish with something like, "We will now entertain your questions." Or the team leader may make a similar statement and field the questions. Either way, the whole team should stand to answer them. As a team, you must plan and practice how you want to handle the questions so that you are not blankly staring at one another to see who will respond. If there is a clear division of expertise, the leader can say, "Since so and so worked on that we'll refer it to her." Or the expert may step forward and say, "I'd like to answer that."

### 7.5.2 Answering Questions

Planting a question or two in the audience may help you get started and also bring up points that you did not want to cover in the formal presentation, but may have time to cover afterwards. Questions are desirable since they indicate interest and may also be enlightening as to further ideas to explore. Always repeat the question for the whole audience to hear and to ensure that you have understood the question. Treat all questioners respectfully. However, if you need clarification, do not hesitate to ask for it. But if you do not know an answer, admit it. The response may be, "That's an interesting question but we did not consider it because of our constraints. We have the data (or the source for it) and will mail it to you immediately." Be sure you do, if you say that. Do not try to flip through slides to hunt for answers unless you are absolutely certain you can find them quickly. It is better to have duplicate slides available at the end if you anticipate having to refer to a particular chart or drawing.

Occasionally you may have troublesome questioners. Maybe they tried to interrupt during the presentation, or maybe they have asked a question that is truly beyond the range of your project. They may simply keep asking questions when others would like to ask about something else. If it is necessary to respond to such a person, remain polite but firmly say, "We'd be happy to discuss that with you personally after the presentation has concluded." You might remind them that you have a limited time in which to cover the required material or that other questioners want to be recognized.

### 7.5.3 Ending

When no more questions are forthcoming, the team leader should conclude the presentation. Do not wait until everyone looks at everyone else and wonders what to do. Simply say, "I see that our time is over (or that there are no more questions). Thank you for your attention." Remember that you are in command of the stage. A brief statement of finality alerts the audience that it is time to applaud. If you planned, practiced, and presented well, they will.

## REFERENCES

Adler, M. (1940). *How To Read A Book.* New York: Simon & Schuster.

Gorman, M., et al. (2001). Transforming the engineering curriculum: lessons learned from a summer at Boeing, *Journal of Engineering Education*, January, 143–149.

Parker, I. (2001). Absolute PowerPoint, *The New Yorker,* May 28, 76–87.

## FURTHER READING

Dominick, P., et al. (2001). *Tools and Tactics of Design,* New York: John Wiley & Sons.

As its name implies, this is a practical text on design written by engineering faculty from a coalition of schools.

Markel, M. (2004). *Technical Communication: Situations and Strategies,* 7th ed., New York: Bedford/St. Martin's Press.

Markel taught at Drexel University in Philadelphia, PA, for several years and his advice is especially apt for senior design. Since he has also consulted with major corporations to improve employees' technical communication skills in the workplace, he knows the industry expectations. His book is now in its seventh edition, but any edition would be helpful.

Pfeiffer, W. (2000). *Technical Writing: A Practical Approach,* Upper Saddle River, NJ: Prentice Hall.

Pfeiffer has a very detailed chapter on graphics, with examples of readability based on font, size, color, and layout, should you want that much information.

Plimpton, G. (n.d.). "How to Make a Speech." Earth and Planetary Sciences 490, Geologic Presentations. http://epswww.unm.edu/facstaff/jgeiss/eps490/howto-make-a-speech.htm (accessed July 19, 2004).

The complete text of George Plimpton's famous advice on "how to make a speech" can be found at this site hosted by the Department of Earth and Planetary Sciences at the University of New Mexico.

Tufte, E.R. (1983). *The Visual Display of Quantitative Information,* Cheshire, CT: Graphics Press.

This landmark text is a compendium of the best in graphics display. Its focus is on impact rather than on which technology was used to create the visual, so it remains relevant. Tufte has more recent books on the topic that are also useful.

- Intellectual property is developed from creations of the mind.
- Protection of intellectual property should begin at the start of your project.
- Keep an engineering notebook, not only to assist you in developing the project, but also to support evidence of invention and to document the details of future patent applications.
- There is a significant difference between trade secrets, patents, trademarks, and copyrights. The cost of certain intellectual property protection can be substantial, so you should carefully consider which is necessary to protect your innovative project and know what you are striving to achieve when discussing the various alternative protections.
- If you are working on your project for a company or other organization, or have a formal customer, you must be aware of who owns which intellectual property. You should have agreements set in place to cover ownership, compensation, and any rights that flow from protecting your intellectual property. Those agreements are in place to protect both you and the organization.
- Licensing intellectual property is one way of gaining monetary benefit from your intellectual property. This process can be complicated and generally requires the assistance of experienced counsel.

# Chapter 8

# INTELLECTUAL PROPERTY*

*Kimberly S. Chotkowski*

## 8.1 INTRODUCTION

*Intellectual property* (*IP*), as differentiated from tangible or real property, refers to creations of the mind. IP rights are provided to the creators of IP to give them a right of limited duration (that is, a limited monopoly) to exclude others from making, using, or selling their creations. In most cases, this limited monopoly is provided in exchange for publicly disclosing their creation and allowing the creation to be made, used, or sold by anyone at the end of the limited monopoly. Publicly disclosing the creation also allows others to create and develop based, in part, on the publicly disclosed creation. IP rights relate to "inventions, literary and artistic works, symbols, names, images, and designs used in commerce" (World Intellectual Property Organization) and allow creators to reveal (or, in the case of trade secrets, to not reveal) their ideas and work, while affording them the opportunity to protect that work from others.

Generally, there are four types of intellectual property: trade secrets, patents, copyrights, and trademarks. A trade secret may be any concrete information (including customer lists, formulas, compilations of information or data, programs, devices, and processes) that is maintained in secret, thus allowing the owner to be the sole user of the information. A classic example of a trade secret is the formula for Coke. Patent laws protect new, useful, and nonobvious discoveries, that is, inventions. Copyright law protects original works of authorship fixed in any tangible medium of expression, such as books, computer code, and paintings. Trademarks

---

\* The information in the chapter is only an introduction to intellectual property laws and a subset of business activities relating to those laws. It is not, nor is it intended to be, legal advice in lieu of appropriate legal counsel. While you can file certain papers discussed in this chapter without an attorney, it is the advice of the author to seek legal assistance when attempting to protect any intellectual property.

and service marks protect marks that distinguish one manufacturer's, merchant's, or service provider's goods or services (for instance, a name or emblem that represents a company or product) from those of others. Below is a discussion of each of these types of IP and what they might mean for you in terms of a design project.

## 8.2 TRADE SECRET

Something is regarded as a trade secret only if the information expected to be protected, such as a recipe for a soft drink or a process for manufacturing a product, is secret and is maintained secret. Once the information is released to the public, protection is no longer afforded to the information; therefore, there is no set time limit on enforcement of a trade secret. To qualify for trade secret protection, there must be some economic benefit to the owner for not divulging the information to the public or ensuring that it is not easily discernable by the public. Finally, there must be a reasonable effort on the part of the owner to maintain secrecy.

A trade secret is not obtained through a formal process; it is merely obtained through the ability of the owner of the trade secret to maintain its secrecy. The Uniform Trade Secret Act enables the owner of the secret information to prevent unauthorized disclosure of the information. The act also allows the owner to prevent dissemination or use of information that has inappropriately been obtained by the person releasing the information to the public or that has been obtained through a confidential relationship. (See confidentiality agreements, which are often referred to as nondisclosure agreements.) In order to bring the body of law to a common ground, the Uniform Trade Secret Act provides guidance to the states; however, not all states have adopted this act, and the protection varies by state.

The protections afforded to trade secret law are much more limited than those of patent law; therefore, the difference between these must be understood before choosing to protect IP through one rather than the other. For instance, you must weigh the cost of obtaining and protecting a patent with the cost of protecting secret information. You might consider whether the timing is correct for obtaining a patent. Also, you might weigh the possibility of your product being reverse-engineered. By requirement, a patent discloses the invention to the public and must enable a competitor to recreate the invention (35 USC § 112). Therefore, if it is impossible to detect your invention through reverse engineering, you may choose to keep it as a trade secret rather than disclosing it as a patent. However, remember that once the secret is revealed you would most likely lose the ability to apply for patent protection. Filing a patent application and prosecuting through to a patent grant can be costly, and, therefore, funds may dictate the attempt to protect your technical solution as a trade secret. Finally, there is a limited period of time in which an inventor can apply for a patent (35 USC § 102); once this period has past, you may be limited to only trade secret protection (35 USC § 102). In some instances, such as the development of software, copyright protection may be a viable alternative.

Typically, trade secret infringement occurs when an employee steals a customer list and then sells it to a competing firm, when an employee leaves one company with proprietary and confidential information and uses that information to the benefit of the new employer, or when a business partner who learns confidential information through a formal agreement later uses this information in an unauthorized manner. The owner of a trade secret can choose to prosecute the infringer; however, significant resources may be needed for doing so.

# 8.3 PATENTS

## 8.3.1 History

The U.S. legal system has roots in the English legal system (Evans v. Eaton 16 US 454,519 [1818]). For instance, the U.S. patent system can trace its origin to a patent granted by Henry VI in 1449 to a Flemish man for a 20-year monopoly on the manufacture of stained glass. These "letters patent" granted monopolies to people who were favored by the King and his supporters. This form of grant became a concern when it was increasingly subject to abuse as more and more patents were granted for all sorts of known (and even staple) goods. After a public outcry, James I was forced to revoke all existing monopolies and declare that grants were only to be used for "projects of new invention." This was incorporated into the Statute of Monopolies in 1623. According to Campbell (1891), in the reign of Queen Anne the rules were changed again so that a written description of the article was given. Over time this written description has evolved into the patent application of today.

The English patent system was carried to the New World by the founders and framers of the U.S. Constitution, who recognized the importance a patent system has to prosperity for a country. This concept was so important to the founders and framers that they wrote the provisions for this system into the Constitution (Article I, Section 8, Clause 8), which provides that the Congress shall have the power "… to promote the Progress of Science and useful Arts, by securing for limited Times to Authors and Inventors and the exclusive Rights to their respective Writings and Discoveries."

The Patent Act was signed into effect by George Washington on April 10, 1790, and was overseen by the Secretary of State, Thomas Jefferson. The first U.S. patent (Figure 8.1) was a handwritten patent awarded in July 1790 to Samuel Hopkins of Pittsford, Vermont, for a method of making potash and pearl ash, a crude form of potassium carbonate, which is used for fertilizer and was an important resource in the agricultural society of the New World.

**FIGURE 8.1**
The first patent.

The patent revealed a method to burn wood ashes a second time before dissolving them to extract potash.

While the Patent Act of 1790 established a patent system, the Patent Office became a separate bureau within the Department of State in 1802. The Patent Office remained in the Department of State until 1849, when it was transferred to the Department of Interior. In 1925 the United States Patent and Trademark Office (USPTO) was transferred to the U.S. Department of Commerce, where it currently resides (35 USC § 1).

The role of the USPTO is to grant patents for the protection of inventions and to register trademarks. It strives to protect and address the concerns and positions of inventors and business entities with respect to their inventions, products, and service marks. The USPTO also offers guidance to a variety of federal offices, including the President of the United States, the Secretary of Commerce, and the Department of Commerce, in all matters involving IP. However, the primary functions of the USPTO, as far as inventors are concerned, are its numerous duties with respect to patents for which it:

- Examines applications and grants patents on inventions when applicants are entitled to them
- Publishes and disseminates patent information
- Records assignments of patents (it maintains search files of U.S. and foreign patents)
- Maintains a search room for public use in examining issued patents and records
- Supplies copies of patents and official records to the public
- Provides training to practitioners and their applicants as to requirements of the patent statutes and regulations
- Publishes the *Manual of Patent Examining Procedure* to elucidate these (see www.uspto.gov)

Similar functions are performed relating to trademarks. By protecting intellectual endeavors and encouraging technological progress, the USPTO seeks to preserve the U.S.'s technological edge, the key to current and future competitiveness. The USPTO also disseminates patent and trademark information that promotes an understanding of intellectual property protection and facilitates the development and sharing of new technologies worldwide (www.uspto.gov).

The USPTO divides the work of examining applications for patents among a number of examining technology centers, each of which is responsible for a certain field. The examiners review applications for patents and determine whether patents can be granted. If the applicant does not agree with the decision of a patent examiner (for instance, when an application is rejected), then an appeal can be taken to a Board of Patent Appeals and Interferences.

At present, the USPTO has over 6,000 employees, of whom about half are examiners. Patent applications are received at the rate of over 300,000 per year. For example, in 2003, the USPTO received 333,452 patent applications and awarded 173,072 patents (for more USPTO statistics, visit their web site at www.uspto.gov).

### 8.3.2 Patent Laws and Practice

The major documents that govern U.S. patent laws, rules, and procedures are Title 35 of the United States Code (USC); Title 37 of the Code of Federal

Register (CFR) and *The Manual for Patent Practice and Procedure* (MPEP). These three documents set forth and define the three statutory categories of patents: utility, plant, and design, which are discussed below. The patent system is a federal system; therefore, both registration of a patent and legal action brought against someone who is making, using, or selling the claimed invention of a patent (a process commonly called *patent infringement*) are governed by federal laws.

### 8.3.3 Types of Patents

#### *Utility Patents*

Generally, when most people think of patents, they are referring to utility patents. Utility patents are awarded for "new, useful, process, machine, manufacture, or composition of matter" (35 USC § 101), which include principles related to the chemical, electrical, and mechanical arts, so long as they meet the statutory requirements as set forth in the Patent Act (35 USC). The word *process* is defined by law as a process, act, or method and primarily includes industrial or technical processes. The term *machine* used in the statute refers commonly to a device consisting of fixed and moving parts that modifies mechanical energy and transmits it in a more useful form. The term *manufacture* refers to articles, which are made, and includes all manufactured articles. The term *composition of matter* relates to chemical compositions and may include mixtures of ingredients as well as new chemical compounds. These classes of subject matter taken together include practically everything that is made by man and the processes for making the products.

As all bodies of law evolve over time, so do patent laws. While there have been many significant changes over the past decade, recent agreements create (for U.S. patent applicants) the opportunity to establish a date of invention outside the U.S. for the purpose of obtaining a patent; create the opportunity to file a new type of patent application called a provisional application; and establish the requirement of a 20-year patent term, (35 USC § 154 [c][1]). Note that the term of a patent is not automatic; rather, it is subject to payment of maintenance fees at 4½, 7½, and 11½ years from the date the patent is granted.

#### *Design Patents*

A design patent is awarded to whoever invents any new, original, and ornamental design for an article of manufacture, again subject to the conditions and requirements of the Patent Act (35 USC § 171). A design patent prohibits a third party from making, selling, or using a product of the protected design. Design patents are purely ornamental and protect no technical elements. For example, to infringe on a design patent directed to a container, the ornamental design of a potentially infringing container and the ornamental appearance shown in the design patent must look alike to the eye of an ordinary observer. It makes no difference what is inside the container or what function is performed by using the two containers. A design patent has a term of 14 years from the date of issuance.

The shape of a product or container can serve as an indicator of the source of the product and therefore its owner or manufacturer, which may be protected with a trademark. The same product shape may also be protected under a design patent. In this case, while the protection afforded by each body of law is unique, the item can obtain protection from both design patent and trademark laws.

## Plant Patents

A person who "invents or discovers and asexually reproduces any distinct and new variety of plant, including cultivated sports, mutants, hybrids, and newly found seedlings, other than a tuber-propagated plant," such as an Irish potato or Jerusalem artichoke, "or a plant found in an uncultivated state," may obtain a patent subject to the conditions and requirements of the Patent Act (35 USC § 161).

An application for a plant patent consists of the same parts as other applications. The term of a plant patent is 20 years from the date on which the patent application was filed in the U.S. or from the earliest filing date of a related patent application.

### 8.3.4   What Is Not Patentable?

Over the history of the patent system the courts have provided guidance for the definition and the limits of the field of subject matter that can be patented. It has been unequivocally held that the laws of nature, physical phenomena, and abstract ideas are not patentable. A patent cannot be obtained upon a mere idea or suggestion. Also, over the years, creators of computer programs, business methods, and even plant-related inventions have fought successful battles to be considered worthy of a patent.

The government has also carved out moral requirements for patents. Patents that relate to illegal acts, such as machines for counterfeiting money, are not patentable. The Atomic Energy Act of 1954 excludes the patenting of inventions useful solely in the utilization of special nuclear material or atomic energy for atomic weapons. Applications filed for national defense are often held under secrecy orders and are treated as a separate group at the USPTO.

### 8.3.5   Who Can File a Patent Application?

The short answer is "anyone." According to the patent statute (35 USC § 1), "Whoever invents or discovers any new and useful process, machine, manufacture, or composition of matter, or any new and useful improvement thereof, may obtain a patent therefore." Thus, any person who is the true inventor, regardless of race, creed, color, or even national origin, may file or cause to be filed a U.S. patent application. An inventor is the person or persons who have created or discovered the information that later becomes the patent. The legal test for inventorship is that a person is an inventor if he or she has invented the subject matter set forth in the claims of a patent application. This implies that anyone who worked on designing the process, machine, etc., is allowed to file a patent, whether or not they were the primary contributor.

### 8.3.6   Who Owns the Patent?

A patent may be owned jointly by two or more inventors (35 USC § 262; MPEP § 301). There does not have to be equal contribution between joint inventors for joint ownership. A joint inventor (and thus a joint owner of a patent application) merely must have contributed to the subject matter set forth in at least one claim of the patent application. Without regard to the other owners, any joint owner of a patent, no matter how small the part interest, may make, use, offer for sale, sell, and import the invention for personal profit, provided he or she does not infringe another's patent rights. Furthermore, a joint owner may sell the interest or any part of it or grant licenses to others without regard to the other joint owners, unless the joint owners have made a contract governing their relation to each other. To avoid unilateral action on the part of one or more joint inventors, it is important to establish, as early as possible and preferably before any patent is granted, the extent of each joint inventor's rights and obligations. An invention may be sold by an inventor to another inventor or another party, in which case the inventor or other party is considered an assignee.

Typically a patent attorney or patent agent who is registered to practice before the USPTO prepares and files a patent application on behalf of an inventor or assignee. However, inventors or assignees of inventions may also file a patent application themselves (MPEP § 401; 37 CFR § 1.31). Engineers working for corporations are typically asked to sign several documents prior to starting their jobs. Frequently, these papers include a nondisclosure agreement (a confidentiality agreement) and an assignment of inventions. In most corporations, these types of documents are standard operating procedure and are nonnegotiable. When a corporation decides to pursue a patent application for a particular invention, the inventor will also be a required to sign a variety of additional documents related to patent procurement process.

When hired by a company in an inventive capacity you receive a salary in exchange for assigning your inventions and the rights that attach to these inventions to your corporation. You will be asked to sign a document, an assignment of inventions, which is a blanket assignment of all of your inventions to the corporation. In short, this means that your company owns your work and is free to apply for patents on it. However, this document should be limited to the scope of work you have been hired to perform. For example, if you are employed as an electrical engineer in the wireless telecommunications business, the assignment document should be limited to such and not be so broad that one can read into it the requirement to assign inventions you have made on your own time relating to sporting equipment. However, when you are at home after regular work hours and awake from a deep sleep with a solution to your corporation's biggest (or even littlest) technical problem, that invention is subject to the assignment document you signed with the company and is the property of the corporation. While it may appear to be unfair that you do not get paid for work you do at home, remember that you are getting paid a regular salary to perform such work. In addition to salaries, many companies have an inventor compensation program, which rewards the efforts of inventors and gives financial incentives to assist in the patent application process.

After inventing and preparing a patent application you will be required to sign documents that accompany the patent application. These declare the following: that you believe yourself to be the original and first inventor, your country of citizenship, and that you either alone or jointly with other inventors are the true inventor of the information contained in the patent application. You will also be asked to sign an individual assignment of invention specific to the invention. Again, this will state that all right, title, and interest in your invention is given to the corporation and that it has the right to prosecute the patent application. In this case *prosecute* refers to the process between the inventor or the assignee and the USPTO, which ultimately results in the patent rejection or grant.

▼

## Example Assignment

**INVENTOR NAME, RESIDENCE**, a citizen of **COUNTRY**, am the inventor of **TITLE OF PATENT APPLICATION** for which the undersigned have executed an application for United States Letters Patent, U.S. Patent Application No. _____, filed _____, 200__.

The undersigned hereby authorize assignee or assignee's representative to insert the Application Number and the filing date of this application if they are unknown at the time of execution of this assignment.

**PRESENT CORPORATION**, a **STATE OF INCORPORATION** corporation, having a principal place of business at **CORPORATE ADDRESS**, (hereafter referred to as the assignee), is desirous of acquiring the entire right, title and interest in said invention, all applications for and all letters patent issued on said invention.

For good and valuable consideration, the receipt and sufficiency of which is acknowledged, the undersigned, intending to be legally bound, do hereby sell, assign and transfer to the assignee and assignee's successors, assigns and legal representatives the entire right, title and interest in said invention and all patent applications thereon, including, but not limited to, the application for United States Letters Patent entitled as above, and all divisions and continuations thereof, and in all letters patent, including all reissues and reexaminations thereof, throughout the world, including the right to claim priority under the Paris Convention or other treaty.

It is agreed that the undersigned shall be legally bound, upon request of the assignee, to supply all information and evidence relating to the making and practice of said invention, to testify in any legal proceeding relating thereto, to execute all instruments proper to patent the invention throughout the world for the benefit of the assignee, and to execute all instruments proper to carry out the intent of this instrument.

The undersigned warrant that the rights and property herein conveyed are free and clear of any encumbrance. **EXECUTED** under seal on this ____ day of _____, 200__ at _____.

(Place)

Witness:

_____    _____(L.S.)

NAME

State of

ss.

County of

On this ____ day of _____, 200__ before me personally appeared **INVENTOR NAME**, to me known to be the person described herein and who executed the foregoing instrument, and acknowledged that he executed the same knowingly and willingly and for the purposes therein contained.

Witness my hand and Notarial seal the day and year immediately above written.

_____

Notary Public

My Commission Expires:

▲

Often all employees are required to sign confidentiality agreements; at minimum those involved in innovative positions will be required to sign them. This document contractually binds you to keep confidential the information you learn and develop while being an employee of the company. It also requires that if your employment terminates with that company, you cannot use the confidential information learned there for the benefit of yourself or a new employer. Confidential information typically includes information such as customer lists, financial information, or the latest corporate technical development. One thing to look for in the confidentiality agreement is that the requirement for confidentiality should not be unreasonably long (3 to 5 years is common). There also should not be an unreasonable limit on your mobility of employment, that is, whether you are excluded from working in the same industry for a long period of time.

The obligations of confidentiality may also be entered into between companies when contemplating joint development of a technical relationship. While you may not be required to sign these documents directly, the obligation to keep confidential what you have learned during these technical and business meetings is imputed to you as a representative or agent of the company. Information that may be considered confidential includes any proprietary information, such as technical solutions and directions, business plans and partners, financial forecasts, and suppliers. As an individual or a design team you may want to bring your ideas or invention to a third party, such as a vendor, a potential financial backer, or an external advisor. In order to preserve the confidential nature of the material and to prevent others from stealing your ideas you should ask each individual to enter into a confidential agreement prior to discussing those topics. See Section 8.6 on intellectual property licensing.

### 8.3.7 Provisional vs. Nonprovisional Patent Applications

Utility and plant patents support two types of applications, a provisional application and a nonprovisional application, whereas a design patent application can only be filed as a nonprovisional application. It is important to understand the difference between provisional and nonprovisional applications if you want to quickly protect your rights.

A provisional application for a patent is a U.S. national application for a patent and is filed in the U.S. Patent and Trademark Office (35 USC § 111[b]; MPEP § 601). It allows filing without a formal patent claim, oath or declaration, or certain other requirements that are necessary for a nonprovisional application (MPEP § 601). It provides the means to establish an early effective filing date for a later nonprovisional patent application, which claims priority over the provisional application. It also allows the term "patent pending" to be applied to any product described in the provisional patent application. Many technology-oriented companies use this type of application for several reasons. The primary reason is that technology development is moving at such a fast pace that by the time an inventor is able to complete the rigorous requirements of the nonprovisional application, someone else may be filing in the U.S. or foreign patent offices. In the U.S., the patent is awarded to the first to invent, so the date is only an indicator of invention, while in most foreign countries the first person to file the application is the owner.

A provisional patent application never results in an enforceable patent; therefore, in order to benefit from a provisional application's earlier filing date an applicant who files a provisional application must file a corresponding nonprovisional application for patent during the 12-month pendency period of the provisional application. If a nonprovisional application is filed within 12 months of the date the provisional application was filed, the provisional application is deemed abandoned and cannot be revived (37 CFR § 1.53[c]; MPEP § 601). There are also some cautions that should be fully investigated when filing a provisional application; namely, provisional applications are not examined on their merits, provisional applications cannot claim the benefit of a previously filed application (either foreign or domestic), and the disclosure of the invention in the provisional application must meet the enablement requirements of the U.S. code (35 USC § 112).

However, a nonprovisional utility patent application can result in a granted patent. A complete nonprovisional utility patent application should contain a specification, which includes the background of the invention, a brief summary of the invention, a brief description of several views of the drawings, a detailed description of the invention, a claim or claims, and an abstract of the disclosure. Drawings should be provided when necessary. The application should also be accompanied by an oath or declaration and the prescribed filing fee (MPEP § 601, 608.01; 37 CFR § 1.71). These requirements are discussed further in Section 8.3.12. In the area of biochemistry, nucleotide or amino acid sequence listings may also be required, when appropriate.

Publication of nonprovisional patent applications is required for most plant and utility patent applications filed on or after November 29, 2000. An applicant may request that the application not be published, so that it is not revealed to competitors, but can only do so if the invention has not been and will not be the subject of an application filed in a foreign country that requires publication 18 months after filing (or earlier claimed priority date) or filed under the Patent Cooperation Treaty (MPEP § 102; 35 USC § 122). Publication occurs after an 18-month period following the earliest effective filing date or priority date claimed by an application.

### 8.3.8 First to File vs. First to Invent

The U.S. system is a "first to invent" system. This means the person who is the first true inventor is ultimately the person who is awarded the patent. Therefore, it should not matter who files first, but rather who invented first. Thus, if two people apply for a patent on the same invention within one year of each other, the true inventor must be determined. However, there are certain presumptions and advantages associated with being the first inventor to actually file the patent application with the USPTO.

In contrast to the U.S., most patent systems in other countries are "first to file" systems. This means that regardless of who invented first, the person who wins the race to the patent office will potentially receive the patent award. The provisional patent allows the inventor to bypass the immediate need for rigorous patent application preparation and receive the earliest filing date possible. The only requirement for a provisional patent is that it contain a written description of the invention (35 USC § 112 ¶1) and any drawings necessary to understand the invention (35 USC § 113).

### 8.3.9 When Should I Begin to Think about Patenting My Invention?

The minute you embark on solving a technical problem is the moment that you should be considering the implications of the patenting process. Maintaining engineering logs or notebooks of your work can be of assistance, not only when establishing a date of invention, but also for collecting detailed information to assist a patent attorney. If you keep an engineering notebook, you should have each entry signed and witnessed by a second person in order to establish any proof of invention.

**159**

CHAPTER 8
Intellectual Property

## 8.3.10 How Do I Know My Invention Is Patentable?

The ultimate determination of patentability is made by the patent examiners and is based on the patent laws and previous patents. However, there are some practical guidelines and some suggestions that may help you answer the question. In the U.S. the test of patentability is based upon the knowledge of an imaginary person who is of ordinary skill in the art (35 USC § 103). This means that if this person would find your invention new, useful, and nonobvious, then it is likely you would be awarded a patent application for that invention.

In order for an invention to be patentable it must be new — as defined in the patent law (35 USC § 102a–g). For example, an invention cannot be patented if "(a) the invention was known or used by others in this country, or patented or described in a printed publication in this or a foreign country, before the invention thereof by the applicant for patent," or "(b) the invention was patented or described in a printed publication in this or a foreign country or in public use or on sale in this country more than one year prior to the application for patent in the United States. ..." These sections of the U.S. Code have been subject to extensive litigation and have taken on specific and special meaning.

If the subject matter of the patent application is able to survive the test as set forth in Section 102, there is still the test of 35 USC § 103, which governs the question of obviousness. Even if the invention is not exactly shown in previously available material and involves one or more differences over the most nearly similar thing already known, a patent may still be refused if the differences would be obvious. The invention must be sufficiently different from what has been used or described before that it may be said to be nonobvious to a person having ordinary skill in the area of technology related to the invention.

So, who is this person of ordinary skill in the art? It is typically a person having a bachelors degree or some other expertise in the field of the invention. The standard is dictated by the complexity of the invention. Just by sheer understanding of the technical problems and solutions present in your field, you will have a baseline idea of what is new and what has been done before. Therefore, you can determine what is new. You are also likely to have the opportunity to confer with trusted colleagues who have either more or different experiences; they will have knowledge that will also help you to determine a baseline level of new and inventive. If you have joined a technical company, there will most likely be an organization responsible for determining the patent applications that will be pursued by the company. If you have such an organization, you should inquire with them immediately for the corporate guidelines.

If you are not with a company and do not have access to a trusted colleague, yet you are interested in learning about the state of the technology in a field, you may wish to perform a patent search. There are several ways to do this. You can go directly to the USPTO in Virginia or you can use their Web site (www.uspto.gov). This site will allow you to search all U.S. patents issued after 1978. The World Intellectual Property Organization also has a Web site (www.wipo.org) that will provide patent applications throughout the world back to 1997. Additionally, you may learn about the state of the technical community in which you are involved by joining a technical organization such as the Institute of Electrical and Electronics Engineers (IEEE). Technical journals and magazines are also a great way to keep up on what is happening in your field. If, after taking a look at these resources, you still have not found your product, then you may have a patentable invention.

### 8.3.11 Patent Fees

The patent process is a notoriously expensive and time-consuming process. Filing fees and other charges range from tens of dollars to over a thousand dollars and can rapidly accumulate (see www.uspto.gov for a list of fees). While you have the right as an individual inventor to file your own patent application, unless you are intimately familiar with the process you will come upon many stumbling blocks. Although the fees associated with the hiring of a patent agent or attorney may be anywhere from $3,000 to over $20,000, depending upon the technical field and the complexity of the invention, we suggest that you hire one. In order to practice as either an attorney or an agent before the USPTO, a person must pass the Examination for Registration Before the United States Patent and Trademark Office, otherwise known as the Patent Bar. For patent attorneys this exam allows the attorney to draft the legal documents that accompany the patent application; an agent is not permitted to perform these legal tasks. Registered patent attorneys or patent agents can be found on the USPTO Web site. This list is kept up to date and is probably the best resource for finding legal representation. However, before retaining an attorney you should consider at least a few questions:

- If you are successful in obtaining a patent, what is the potential market and what if any is the value you may target for a return?
- Is there more information required before the invention can be performed? That is, are you ready to file the patent?
- Even if you successfully obtain a patent, is it so imbedded in a product that you will not know if someone is infringing on the patent? Should you instead keep the technology a trade secret?
- Are there underlying patents that you may need to gain a license in order to practice your invention?
- Is this a monumental step in the technology? That is, is it a pioneering patent or is it a small improvement to existing technology?

These types of questions will help you place a value on the invention and determine the amount of money that you are willing to spend. Corporate entities consider these same questions as well as others, such as whether the invention is aligned with the corporate business and technical plan.

### 8.3.12 What Should Be in a Patent Application?

Legally, the requirements for a complete application include the filing of a specification, a drawing, an oath by the applicant, and submitting a fee. A patent application should also have other documents, including a transmittal form, an associated form for the fee, and a return receipt postcard. However, the heart of the application is the specification, which should have the following ordered sections:

1. Title of the invention. (37 CFR § 1.72; MPEP § 608.01[b])

2. Cross-reference to related applications (if any). (Related applications may be listed on an application data sheet, either instead of or together with being listed in the specification.) (35 USC § 119[c], 120)
3. Statement of federally sponsored research and development (if any).
4. Reference to a "sequence listing," a table, or a computer program–listing appendix submitted on a compact disc and incorporation by reference of the material on the compact disc, including duplicates.
5. Background of the invention. (MPEP § 608.01[c])
6. Brief summary of the invention. (MPEP § 608.01[a])
7. Brief description of the several views of the drawing (if any). (MPEP § 608.01[f]; 37 CFR § 1.74)
8. Detailed description of the invention. (MPEP § 608.01[g]; 37 CFR § 1.71)
9. Claim or claims. (MPEP § 608.01[i]; 37 CFR § 1.75)
10. Abstract of the disclosure. (37 CFR § 1.72[b]; MPEP § 608.01[b])
11. Sequence listing (if any).
12. Drawings. (MPEP § 608.02; 35 USC § 113)

We will briefly step through each of these sections to show the level of detail involved in a patent application (see MPEP § 601, 608.01[a]). The title (1) of the invention should be as short and specific as possible (no more than 500 characters). It should appear as a heading on the first page of the specification, if it does not otherwise appear at the beginning of the application. A cross-reference to related or other applications enables the patent to claim priority to earlier dates of invention, such as to a provisional application. Without such reference, the earliest date may be surrendered, and you may lose the benefits of filing your provisional application. A reference to federally sponsored work (3) and sequence listings (4) are included where appropriate. The background of the invention (5) sets forth the story of your invention and provides the examiner a basis for the invention and the problem that it solves. A brief summary of the invention (6), indicating its nature and substance, which may include a statement of the object of the invention, should precede the detailed description. The summary should be commensurate with the invention's broadest claim, and any object discussed should be the most general overview. Drawings (12) are required if they are necessary to the understanding of the invention; it is rare that drawings are not included. Typically, a drawing is required for each element of the claim. The specification includes a brief description of the drawings (7) in order to orient the readers as to what they are actually viewing.

The detailed description (8) of the invention should refer to the different views by specifying the numbers of the figures and to the different parts by use of reference numerals. The description of the invention must set forth the manner and process of making and using it and is required to be in such full, clear, concise, and exact terms as to enable any person skilled in the technological area to which the invention pertains, or with which it is most nearly connected, to make and use it.

The specification, that is, the description, must set forth the precise invention for which a patent is solicited, in such manner as to distinguish it from other inventions and from what is old. It must completely describe a specific embodiment of the process, machine, manufacture, composition of matter, or improvement invented and must explain the mode of operation or principle whenever applicable. The best mode contemplated by the inventor for carrying out the invention must be described.

In the case of an improvement, the specification must particularly point out the part or parts of the process, machine, manufacture, or composition of matter to which the improvement relates. The description should be confined to the specific improvement and to such parts as necessarily cooperate with it or may be necessary for complete understanding or description of it.

The specification must conclude with a claim or claims (9) particularly pointing out and distinctly claiming the subject matter that the applicant regards as the invention. The portion of the application in which the applicant sets forth the claim or claims is arguably the most important part of the application, as it is the claims that define the scope of the protection afforded by the patent and by which questions of infringement are judged by the courts. It is the claims that rely on the fundamental art of a patent attorney and must be drafted with care.

A brief abstract of the technical disclosure (10) of the specification, including that which is new in the art to which the invention pertains, must be set forth on a separate page immediately following the claims. The abstract should be in the form of a single paragraph of 150 words or less, under the heading "Abstract of the Disclosure." Sequence listings (11) must be included if necessary.

The first page of a granted patent is published as shown in Figure 8.2. The numbers on the cover page of the patent represent the following:

[12] The last name of the first listed inventor on the patent.
[10] Patent number.
[45] Date the patent is issued.
[54] Title of the patent.
[75] Inventors' names and places of residence.
[73] Assignees, that is, patent owners, and their places of business.
[21] Application number.
[22] Filing date of the patent application.
[60, 62] Related applications from which the patent is claiming priority.
[51] "International Classification" or "International Patent Classification."
[52] U.S. classification codes assigned by the Patent Office examiner, which provide the search area where similar patents may be found.
[58] U.S. classification codes that the examiner searched for prior art.
[56] References made of record during prosecution of the application. These can be U.S. patents, foreign patents, or other publications. Following the references are the names of the primary examiner at the Patent Office, the assistant examiner (if any), and the attorney, agent or firm of record.
[57] Abstract of the invention.
[74] The attorney or agent representing the inventor during the patenting process.

The cover page concludes with the number of claims and drawings as well as a representative drawing of the patent.

### 8.3.13 Now That I Have a Patent, What Can I Do with It?

A patent is an exclusionary right (MPEP § 301; 35 USC § 154). It is often referred to as a ticket to the courtroom. For a limited time (20 years after the date of filing the patent application), the patent allows you to exclude others from making, using, offering to sell, selling, or importing into the U.S. the

US006215778B1

(12) **United States Patent**
Lomp et al.

(10) Patent No.: **US 6,215,778 B1**
(45) Date of Patent: **Apr. 10, 2001**

(54) **BEARER CHANNEL MODIFICATION SYSTEM FOR A CODE DIVISION MULTIPLE ACCESS (CDMA) COMMUNICATION SYSTEM**

(75) Inventors: **Gary Lomp**, Centerport; **John Kowalski**, Hempstead; **Fatih Ozluturk**, Port Washington; **Avi Silverberg**, Commack; **Robert Regis**, Huntington; **Michael Luddy**, Sea Cliff; **Alexander Marra**, New York; **Alexander Jacques**, Mineola, all of NY (US)

(73) Assignee: **InterDigital Technology Corporation**, Wilmington, DE (US)

( * ) Notice: Subject to any disclaimer, the term of this patent is extended or adjusted under 35 U.S.C. 154(b) by 0 days.

(21) Appl. No.: **08/956,740**

(22) Filed: **Oct. 23, 1997**

**Related U.S. Application Data**

(62) Division of application No. 08/669,775, filed on Jun. 27, 1996, now Pat. No. 5,799,010.

(60) Provisional application No. 60/000,775, filed on Jun. 30, 1995.

(51) Int. Cl.[7] ............................... H04B 7/214

(52) U.S. Cl. ......................... 370/335; 370/465

(58) Field of Search ........................ 370/310, 318, 370/320, 328, 329, 335, 342, 465, 468, 477, 441, 479, 503, 515, 375/140, 146, 147, 354, 367

(56) **References Cited**

U.S. PATENT DOCUMENTS

| | | |
|---|---|---|
| 4,901,307 | 2/1990 | Gilhousen et al. |
| 5,022,049 | 6/1991 | Abrahamson et al. |
| 5,164,900 | 1/1992 | Taylor |

(List continued on next page.)

FOREIGN PATENT DOCUMENTS

| | | |
|---|---|---|
| 374373A3 | 7/1994 | (DE) |

| | | |
|---|---|---|
| 743732A1 | 7/1989 | (DE) |
| 0468839A2 | 1/1992 | (EP) |

(List continued on next page.)

OTHER PUBLICATIONS

Zhao Liu et al., "SIR–Based Call Admission Control for DS–CDMA Cellular Systems", IEEE Journal on Selected Areas in Communications, US, IEEE Inc. New York, vol. 12, No. 4, May 1, 1994, pp. 638–644.

(List continued on next page.)

*Primary Examiner*—Wellington Chin
*Assistant Examiner*—Kwang B. Yao
(74) *Attorney, Agent, or Firm*—Volpe and Koenig, P.C.

(57) **ABSTRACT**

A multiple access, spread-spectrum communication system processes a plurality of information signals received by a Radio Carrier Station (RCS) over telecommunication lines for simultaneous transmission over a radio frequency (RF) channel as a code-division-multiplexed (CDM) signal to a group of Subscriber Units (SUs). The RCS receives a call request signal that corresponds to a telecommunication line information signal, and a user identification signal that identifies a user to receive the call. The RCS includes a plurality of Code Division Multiple Access (CDMA) modems, one of which provides a global pilot code signal. The modems provide message code signals synchronized to the global pilot signal. Each modem combines an information signal with a message code signal to provide a CDM processed signal. The RCS includes a system channel controller is coupled to receive a remote call. An RF transmitter is connected to all of the modems to combine the CDM processed signals with the global pilot code signal to generate a CDM signal. The RF transmitter also modulates a carrier signal with the CDM signal and transmits the modulated carrier signal through an RF communication channel to the SUs. Each SU includes a CDMA modem which is also synchronized to the global pilot signal. The CDMA modem despreads the CDM signal and provides a despread information signal to the user. The system includes a closed loop power control system for maintaining a minimum system transmit power level for the RCS and the SUs, and system capacity management for maintaining a maximum number of active SUs for improved system performance.

**6 Claims, 37 Drawing Sheets**

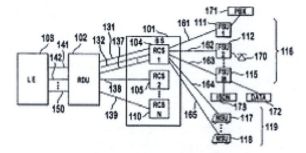

FIGURE 8.2
The first page of a patent.

subject matter that is within the scope of protection granted by the patent. It is not a right or assurance that you will make money from the grant of a patent.

Patent litigation can cost tens of millions to hundreds of millions of dollars; therefore, this is not something that is easily completed by an individual. However, there have been several accounts of individual inventors successfully defeating large corporate entities. One such is that of Robert W. Kearns, whose invention came about because his eyesight was compromised when, in 1953, he popped open a bottle of champagne on his honeymoon. He experienced difficulty seeing through a two-speed windshield wiper and so developed an intermittent wiper.

Engineers at car companies around the world were also working on this idea, but it was Kearns who was awarded the patent as the first true inventor. The patent was granted in 1967; however, it took until 1990, when he successfully settled his suit with Ford Motor Company, for him to win any money through patent infringement lawsuits. He then settled with Chrysler in 1992 and won over $30 million from these suits (See *Kearns v. Chrysler Corp.,* 32 F.3d 1541, 1547, 31 USPQ2d 1746, 1750 [Fed. Cir. 1994]). While these cases do occur, they are few and far between. If litigation is your method of enforcement, be prepared for a long battle (2 to 10 years) and a significant cost.

## 8.4 COPYRIGHTS

The law of copyright protects "original works of authorship" that are fixed in a tangible form of expression. A work must not be a copy from another author; that is, it must be generated by the copyright holder. Unlike a patent, there is no requirement that it be new or even nonobvious from prior works. Copyrightable works include:

- Literary works
- Musical works, including any accompanying words
- Dramatic works, including any accompanying music
- Pantomimes and choreographic works
- Pictorial, graphic, and sculptural works
- Motion pictures and other audiovisual works
- Sound recordings
- Architectural works

These categories should be viewed broadly. For example, computer programs and most "compilations" may be registered as literary works, while maps and architectural plans may be registered as pictorial, graphic, and sculptural works (17 USC § 102).

Several categories of material are generally not eligible for federal copyright protection. These include:

- Works that have *not* been fixed in a tangible form of expression, for example, choreographic works such as ballets or modern dances that are not recorded
- Improvisational speeches or performances that have not been fixed in a tangible medium such as a recording or written record
- Titles, names, short phrases, and slogans
- Familiar symbols or designs
- Mere variations of typographic ornamentation, lettering, or coloring
- Mere listings of ingredients or contents
- Ideas, procedures, methods, systems, processes, concepts, principles, discoveries, or devices (excluding a description, explanation, or illustration of these)
- Works consisting *entirely* of information that is common property and containing no original authorship, for example, rules of measure, telephone numbers, addresses, the periodic table, or other lists or tables taken from public documents or other common sources.

### 8.4.1 Who Can Copyright and Who Owns the Copyright?

Copyright protection attaches at the moment the subject work is created in fixed form; that is, the copyright immediately becomes the property of the author who created the work. Only the author or a group or entity that has received rights from the author can claim copyright. Mere ownership of a book, manuscript, painting, or any other copy or phonorecord does not give an owner the copyright. The law provides that transfer of ownership of any material object that embodies a protected work does not of itself convey any rights in the copyright.

In the case of works made for hire, the employer (not the employee) is considered to be the author. A "work made for hire" is a work prepared by an employee within the scope of his or her employment or a work specially ordered or commissioned for use as: a contribution to a collective work, a part of a motion picture or other audiovisual work, a translation, a supplementary work, a compilation, an instructional text, a test, answer material for a test, or an atlas. A work can be deemed made for hire if the parties expressly agree in a signed written instrument.

The authors of a joint work are co-owners of the copyright in the work, unless there is an agreement to the contrary. For collective works or compilations, a copyright exists in each separate contribution and separately in the collective work as a whole and vests initially with the author of each contribution.

## 8.4.2   What Rights Does a Copyright Afford?

Copyright protection is available to both published and unpublished works. Copyright ownership generally gives the owner of copyright the exclusive right to do and to authorize others to do the following:

- To reproduce the work in copies or phonorecords
- To prepare derivative works based upon the work
- To distribute copies or phonorecords of the work to the public by sale or other transfer of ownership, or by rental, lease, or lending
- To perform the work publicly, in the case of literary, musical, dramatic, and choreographic works, pantomimes, motion pictures, and other audiovisual works
- To display the copyrighted work publicly, in the case of literary, musical, dramatic, and choreographic works, pantomimes, and pictorial, graphic, or sculptural works, including the individual images of a motion picture or other audiovisual work
- To perform the work publicly by means of a digital audio transmission, in the case of sound recordings

It is illegal for anyone to violate any of the rights provided by the copyright law to the owner of copyright. However, there are certain limitations on these rights. In some cases, these limitations are specified exemptions from copyright liability. One major limitation is the doctrine of "fair use." In other instances, the limitation takes the form of a "compulsory license" under which certain limited uses of copyrighted works are permitted upon payment of specified royalties and compliance with statutory conditions.

## 8.4.3   Fair Use

You may have seen fair use signs at your local or school library. Fair use essentially allows the use of some work under some circumstances, without the explicit permission of the copyright holder, that is, fair use is a defense to copyright infringement (17 USC § 107). To establish fair use you must ask:

- What is the character of the use? Uses that are nonprofit, educational, and personal, used as criticism, commentary, news reporting, and parody are fair use, while commercial uses imply infringement.
- What is the nature of the work to be used? Works that contain facts are subject to fair use, while imaginative works or those with original input may be infringed upon.
- How much of the work will you use? A small subset of the work, such as a few pages, but not chapters, is fair use, while more than a small amount is infringement. Many of you probably have had professors copy chapters of texts. This is controversial, but it is most likely that the copying of a complete chapter is not fair use of the copyrighted material.
- What effect would this use have on the market for the original or for permissions if the use were widespread? If the market is not economically impacted, then it is most likely fair use. If the copying takes away the economic benefit to the copyright owner, then it is likely an infringement.

### 8.4.4   How Do I Obtain Copyright Protection?

The way in which copyright protection is secured is frequently misunderstood. No publication or registration or other action in the Copyright Office is required to secure copyright. The copyright automatically attaches upon the moment the work is in a fixed tangible form. However, you should mark your work as being copyrighted by indicating on the medium your name, the date, and the copyright symbol (example: Kimberly S. Chotkowski © 2005). This provides public notice of your ownership interest and may be helpful in enforcement.

You may also register your copyright; it is a legal formality intended to make a public record of a particular copyright. Although registration is not a requirement for protection, the copyright law provides several inducements or advantages to encourage copyright owners to register their copyrights. Among these advantages are the following:

- Registration establishes a public record of the copyright claim. Before an infringement suit may be filed in court, registration is necessary for works of U.S. origin.
- If made before or within 5 years of publication, registration will establish *prima facie* evidence in court of the validity of the copyright and of the facts stated in the certificate. If registration is made within 3 months after publication of the work or prior to an infringement of the work, statutory damages and attorney's fees will be available to the copyright owner in court actions. Otherwise, only an award of actual damages and profits is available to the copyright owner.
- Registration also allows the owner of the copyright to record the registration with the U.S. Customs Service for protection against the importation of infringing copies.
- Registration may be made at any time within the life of the copyright.

### 8.4.5   How Long Does a Copyright Last?

There are several acts that govern the copyright statute and dictate the period of time that a copyright can be enforced.

**FIGURE 8.3**
Examples of trademarks.

A work that is created (fixed in tangible form for the first time) on or after January 1, 1978, is automatically protected from the moment of its creation and is ordinarily given a term enduring for the author's life plus an additional 70 years after the author's death. In the case of "a joint work prepared by two or more authors who did not work for hire," the term lasts for 70 years after the last surviving author's death. For works made for hire, and for anonymous and pseudonymous works (unless the author's identity is revealed in Copyright Office records), the duration of copyright will be 95 years from publication or 120 years from creation, whichever is shorter.

## 8.5 TRADEMARKS AND SERVICE MARKS

A trademark is a word, phrase, symbol, design, or a combination of words, phrases, symbols, or designs, that identifies and distinguishes the source of the goods of one party from those of others (Figure 8.3). This includes scents, color, words, artistic designs, nonfunctional designs, patterns, symbols, and packaging. A service mark is similar to a trademark, but it identifies and distinguishes the source of a service rather than a product. The shape of the product or the package in which the product is marketed, or of an element of the packaging, can also comprise a mark or an element of a mark. This type of trademark, called a configuration mark, is often referred to as "trade dress."

Under trademark law, the mark must be such as to not cause confusion, mistake, or deception among the consuming public. In making this determination the circumstances will take into account what was actually sold under each mark. In addition, courts analyzing a trademark infringement action would look at the buying habits of the consuming public and the sales methods used by each manufacturer.

### 8.5.1 The Symbols TM, SM, and ®

The TM (trademark) or SM (service mark) designation is used to alert the public to the claim of ownership of the mark, regardless of whether you have filed an application with the USPTO. However, you may use the federal registration symbol ® only after the USPTO actually registers a mark, and not while an application is pending.

Trademark rights may be used to prevent others from using a confusingly similar mark but not to prevent others from making the same goods or from selling the same goods or services under a clearly different mark. Trademarks that are used in interstate or foreign commerce may be registered with the Patent and Trademark Office.

To obtain a trademark an application does not have to be filed with the Patent and Trademark Office. Rights in a mark can be based on legitimate use of the mark. These marks are governed by common law. However, owning a federal trademark registration on the Principal Register provides several advantages. Federal registration provides constructive notice to the public of ownership of the mark, a legal presumption of the registrant's ownership of the mark and the registrant's exclusive right to use the mark nationwide on or in connection with the goods or services listed in the registration, the ability to bring an action concerning the mark in federal court, the use of the U.S. registration as a basis to obtain registration in foreign countries, and the ability to file the U.S. registration with the U.S. Customs Service to prevent importation of infringing foreign goods.

A trademark application must include the name of the applicant, a name and address for correspondence, a clear drawing of the mark, a listing of the goods or services, and the filing fee for at least one class of goods or services (37 CFR § 2.21).

### 8.5.2   What Cannot Be Trademarked?

Under federal code (15 USC § 1052), no trademark by which the goods of the applicant may be distinguished from the goods of others shall be refused registration on the principal register on account of its nature unless it includes marks that comprise:

- Immoral, deceptive, or scandalous matter; or matter which may disparage or falsely suggest a connection with persons, living or dead, institutions, beliefs, or national symbols, or bring them into contempt, or disrepute; or a geographical indication which, when used on or in connection with wines or spirits, identifies a place other than the origin of the goods
- Governmental flags, coats of arms or other insignia of any State or municipality, or of any foreign nation, or any simulation thereof
- A name, portrait, or signature identifying a particular living individual … or the name, signature, or portrait of a deceased President of the United States during the life of his widow, if any (except by the written consent of the individual or widow respectively)
- Marks which so resemble a mark registered in the Patent and Trademark Office, or a mark or trade name previously used in the United States by another and not abandoned, as to be likely, when used on or in connection with the goods of the applicant, to cause confusion, or to cause mistake, or to deceive (barring concurrent registration allowances)
- Merely descriptive or deceptively misdescriptive marks, geographically descriptive or misdescriptive marks, surnames, or functional marks

### 8.5.3   How Long Does a Trademark Last?

Trademark registrations issued or renewed on or after November 16, 1989, remain in force for 10 years from their date of issue or the date of renewal and may be renewed for periods of 10 years, with the exception of cancelled or surrendered trademarks.

## 8.6   INTELLECTUAL PROPERTY LICENSING

In addition to filing an infringement action to preserve your exclusive right to a patent, there are other ways to make a profit from your ideas. One such way is through licensing of IP. Through licensing, technology know-how, patents, trademarks, and copyrights can be bought, sold, transferred in whole or in part, and leased. For example, you can permit a company to implement your process for a set amount of time in exchange for a given amount of money. Joint ventures and franchises are other ways to accomplish the exchange of IP rights. In exchange for some value, either monetary or even an exchange of IP, a license grants to another party certain rights to use the identified IP. A license can also be exclusive (limiting the transfer rights to one group) and nonexclusive (transferring to multiple groups) or can be exclusive or nonexclusive based upon a certain field of technical use, geographic locations, or other limits agreed upon by both parties.

IP licensing is a mechanism to obtain monetary reward for the investment of creating. Licensing is a large aspect of corporate business and can provide a means for an individual inventor to benefit from the investment. Patent and trademark licensing is not compulsory or required in the U.S., but it is in many countries.

If you believe your idea may be patentable or that it may turn out to be profitable, then before you approach someone about IP licensing, you should require them to enter into a confidential agreement. You may use the nondisclosure agreement provided at the end of this chapter as an example or you may draft your own agreement with the assistance of an attorney. As discussed previously, this agreement prohibits someone from profiting from your idea after learning it from you. You can also ask vendors or other people who may assist you on the project to sign a confidentiality agreement for the same reasons.

There are significant legal implications related to the licensing of IP; therefore, we suggest that you contact a patent attorney or business attorney if you are interested in licensing your product. However, before you contact an attorney you should have a good understanding of the market, your competitors, and your potential targets. This will help you to work with the attorney who will assess the coverage of the patent to an existing product or to a desired market and assist in the monetary and legal evaluations.

---

▼

---

## SAMPLE CONFIDENTIAL AGREEMENT OR NON-DISCLOSURE AGREEMENT

This AGREEMENT dated as of _____, 20__ ("Effective Date"), by and between _____, a [**STATE of INCORPORATION**] corporation, having offices at _____ (hereinafter referred to as "_____"), and _____, a _____ corporation, having offices at _____ (hereinafter referred to as "_____"). The above companies are collectively referred to as the "Parties" and individually referred to as a "Party."

### W I T N E S S E T H:

WHEREAS the Parties desire to engage in discussions relating to: _____, together with any potential business relationship, which arises between the Parties as a result of such discussions (the "Relationship").

WHEREAS, in furtherance of the Relationship, it is essential that the Parties exchange certain confidential marketing, technical and commercial information under the terms and conditions specified below.

NOW THEREFORE, the Parties, intending to be legally bound, hereby agree as follows:

1. For the purpose of this Agreement, "Confidential Information" means all confidential and/or proprietary information, disclosed by one Party to the other following the Effective Date of this Agreement, and whether oral, electronic, visual or written form, including, without limitation, processes, services, products, plans, intentions, pricing, inventions, intellectual property rights, trade secrets, know-how, methods, techniques, computer software, source and object codes, algorithms, engineering concepts, product specifications, models, descriptions, drawings, samples, demonstrations, manufacturing processes, research and development efforts, development tools, marketing information, sales, suppliers, customers, and financial data.

2. The Parties shall use the other Party's Confidential Information only for the purposes of considering and performing the Relationship. The Parties shall not disclose, or allow disclosure of, such Confidential Information to any third party without the prior written approval of the other Party.

3. Each Party shall protect the confidentiality of such Confidential Information with the same degree of care it uses to protect its own Confidential Information, which measures will, at a minimum, be in accordance with generally accepted business standards for protecting confidential and proprietary business information.

4. Each Party shall limit the distribution and disclosure of such Confidential Information to only the minimum number of its employees or representatives who have a need to know for the purposes of the Relationship. The Party disclosing Confidential Information to its employees or representatives shall (i) ensure that such persons personally adhere to and comply with all terms and obligations of confidentiality, use and protection of the Confidential Information as accepted by the Parties under this Agreement and (ii) be liable if such persons do not adhere to such requirements.

5. The disclosing Party shall use all reasonable efforts to (i) identify the confidential nature of the Confidential Information by proprietary and/or confidential notices and legends and (ii) identify the confidential nature of oral or visually disclosed Confidential Information at the time of disclosure and reduce oral Confidential Information to writing within thirty (30) days of its disclosure, marked as set forth above. Notwithstanding the foregoing, with respect to Confidential Information other than in written form, the disclosing Party need only identify such Confidential Information as confidential once. The disclosing Party shall have the right to provide advance written designation of the confidential nature of the Confidential Information, which will be disclosed in other than written form without the need to follow up in writing. Notwithstanding the foregoing, the failure to mark Confidential Information as confidential or proprietary, or to record oral conversations in writing, said information shall be deemed Confidential Information hereunder if it is of the type or nature that would reasonably be expected by the Parties, in the context of its disclosure, to be confidential.

6. A Party may copy Confidential Information provided by the other Party only to the extent reasonable or necessary for the Relationship.

All copies shall always clearly contain the same proprietary and confidential notices and legends, which appear on the original Confidential Information. Confidential Information shall remain the property of the disclosing Party.

7.  Within thirty (30) days of receipt by the receiving Party of a written request from the disclosing Party, and in any event within thirty (30) days of termination of this Agreement, the Confidential Information together with all copies, summaries, analyses and extracts thereof shall be returned to the disclosing Party or destroyed at the disclosing Party's election, and if requested by the disclosing Party, the receiving Party shall provide clear evidence of such destruction through, by way of example, written certification. The receiving Party may retain an archival copy for use (on a confidential basis) solely in the adjudication of a dispute pertaining to this Agreement

8.  This Agreement shall become effective on the Effective Date stated above and shall remain in effect for a period of two (2) years unless sooner terminated or extended in writing by the Parties.

9.  Notwithstanding the termination or expiration of this Agreement, the confidentiality, use, venue, compliance with export control laws and governing law provisions contained herein shall remain in full force and effect for four (4) years from the date of termination of this Agreement.

10. The obligation of confidentiality shall not apply to the extent the Confidential Information (i) was previously known to the receiving Party free of any obligation to keep it confidential at the time it was communicated by the disclosing Party, or (ii) is or becomes generally known to the public, provided that such public knowledge is not the result of any acts attributable to the receiving Party, or (iii) which the disclosing Party explicitly agrees in writing need not be kept confidential, or (iv) is disclosed pursuant to any judicial or governmental requirement or order, provided that the receiving Party takes reasonable steps to give the disclosing Party sufficient notice in order to contest such requirement or order, or (v) is independently developed by the receiving Party without reliance on the disclosing Party's Confidential Information and that such fact can be demonstrated to the reasonable satisfaction of the disclosing Party, or (vi) is rightfully received by the receiving Party from a third party, free of any obligation of confidentiality.

11. Nothing contained herein shall be construed as conferring, by implication, estoppel or otherwise, any license or right in respect of any trademark or any copyright or other intellectual property right or any Confidential Information or any invention or any existing or later issued patent. No other rights or obligations other than those expressly recited herein are to be implied from this Agreement.

12. This Agreement shall be construed in accordance with and governed by Pennsylvania law without reference to conflict of laws principles. The Parties irrevocably consent to the venue and jurisdiction of the state and federal courts in the Commonwealth of Pennsylvania for the purposes of enforcing the provisions herein. Process may be served on either Party by U.S. Mail, postage pre-paid, certified or registered, or globally recognized overnight mail service, and addressed to the attention of the respective Parties' General Counsel at the address as indicated on page 1 of this Agreement.

13. The Parties agree that the Confidential Information furnished hereunder is of a unique nature and of extraordinary value and of such character that any unauthorized use or disclosure thereof by the recipient will cause irreparable injury to the disclosing Party for which the disclosing Party will have no adequate remedy at law. Accordingly, in the event of actual or threatened unauthorized use or disclosure, the disclosing Party shall have the right, in addition to all other remedies at law or in equity, to have the provisions of this Agreement specifically enforced by any court having equity jurisdiction and to seek a temporary or permanent injunction or order prohibiting the recipient, its agents, officers, directors, and employees, as the case may be, from such unauthorized use or disclosure of any Confidential Information provided pursuant to this Agreement. In any proceeding by the disclosing Party to obtain injunctive relief, the receiving Party's or any other defendant's ability to answer in damages shall not be a bar or be interposed as a defense to the granting of relief and the disclosing Party shall not be required to post a bond or other undertaking in such a proceeding.

14. Both Parties will adhere to all applicable laws, regulations and rules relating to the export and re-export of technical data and agree not to transfer any Confidential Information received hereunder to any country prohibited from obtaining such data according to any national export regulation (e.g., U.S. Department of Commerce Export Administration Regulations) without first obtaining all valid export licenses and authorizations.

15. Unless otherwise stated _____ and _____, as used herein, shall include all affiliates (specifically, any other entity which controls, is controlled by or is under common control of a Party), as applicable, of each such defined entity.

16. Neither Party may reverse engineer, reverse assemble or de-compile any part of the other Party's Confidential Information without first obtaining the other Party's written consent.

17. If any provision of this Agreement is held invalid or unenforceable by a competent court, it is the Parties' intent that the remaining provisions shall be in full force and effect.

18. This Agreement contains the entire agreement of and supersedes any and all prior understandings, arrangements and agreements between the Parties hereto whether oral or written, with respect to the subject matter hereto.

19. Neither Party shall be entitled to assign or transfer this Agreement nor any rights or obligations contained herein to any third party without the prior written approval of the other Party hereto.

20. This Agreement may only be amended in a writing signed by both Parties.

21. This Agreement may be executed by the Parties in two (2) original counterparts.

\*      \*      \*

IN WITNESS WHEREOF, the Parties hereto have caused this Non-Disclosure Agreement to be executed by their duly authorized officers as of the Effective Date above.

_____

(Company Name) or (Individual Name)

By: _____

Name:

Title:

Legal Dept. Signature: _____

_____

(Company Name) or (Individual Name)

By: _____

Name:

Title:

EXECUTED under seal on this _____ day of _____, 20__ at _____
     (Place)

Witness:

_____          _____(L.S.)

▬▬▬▬▬▬▬▬▬▲▬▬▬▬▬▬▬▬

## REFERENCES

Campbell, L.H. (1891). _The Patent System of the United States so Far as It Relates to the Granting of Patents._ Washington, DC: Press of McGill and Wallace.

An interesting early history of the U.S. patent and trademark office and process.

Donner, I.H. (2003). _Patent Prosecution Practice and Procedure Before the US Patent Office,_ 3rd ed. Washington, DC: Bureau of National Affairs.

A thorough treatise on intellectual property. It includes court cases and a detailed summary of intellectual property law. This text is used as a reference by many patent attorneys.

Mayers, H.R. and Brunsvold, B.G. (1995). *Drafting Patent License Agreements,* 3rd ed. Washington, DC: Bureau of National Affairs.

> A guide for a variety of different types of license clauses and agreements. This text includes many examples of license agreements that can be used as a reference for understanding forms of licensing.

Title 17 of the U.S. Code http://uscode.house.gov (accessed July 19, 2004).

> Title 17 of the USC covers copyright laws.

Title 35 of the U.S. Code http://uscode.house.gov (accessed July 19, 2004).

> The U.S. Code is the compendium of the laws of the United States of America. In particular, Title 35 of the U.S. Code is the set of laws governing patents and trademarks.

Title 37 of the Code of Federal Regulations http://www.gpoaccess.gov/cfr (accessed August 30, 2004).

> Title 37 of the Code of Federal Regulations is the set of rules governing the practices of intellectual property.

U.S. Department of Commerce Patent and Trademark Office. (August 2001). *Manual of Patent Examining Procedure,* 8th ed. U.S. Department of Commerce Patent and Trademark Office.

> This comprehensive text contains all U.S. intellectual property rules and procedures as well as the USPTO's application filing requirements. This is the publication used by patent agents and attorneys when filing and prosecuting applications. It also includes a detailed chapter on the international patent filing requirements.

www.copyright.gov (accessed August 30, 2004).

> The Copyright Office performs the same functions for copyrights as the USPTO does for patents and trademarks.

www.uspto.gov (accessed July 19, 2004).

> The USPTO is the federal agency responsible for administering the laws related to patents and trademarks. From the USPTO Web site you can find information about these laws and how to apply for patents and trademarks and conduct searches for existing patents and trademarks.

www.wipo.int (accessed August 30, 2004).

> The World Intellectual Property Organization is an agency of the United Nations that administers international treaties related to IP protection. From the Web site you can search recent international patents and obtain information about the international patent process.

## FURTHER READING

Grubb, P.W. (1999). *Patents for Chemicals, Pharmaceuticals and Biotechnology: Fundamentals of Global Law, Practice and Strategy,* 3rd ed. London/New York: Oxford University Press.

> This guide provides specific examples and requirements for patenting in the fields of chemicals, pharmaceuticals, and biotechnology.

The Inventors Assistance Center (IAC) provides patent information and services to the public. The IAC is staffed by former supervisory patent examiners and experienced primary examiners who answer general questions concerning patent examining policy and procedure. For contact information, see http://www.uspto.gov/web/offices/pac/dapp/pacmain.html.

Miller, A.R. and Davis, M.H. (2000). *Intellectual Property: Patents, Trademarks, and Copyright*, 3rd ed. St. Paul, MN: West Group Publishing.

This is well-written and readable pocket guide that is a basic overview of all areas of intellectual property and is useful for those who need an introduction to the field.

Pressman, D. (2002). *Patent It Yourself,* 9th ed. Berkeley, CA: Nolo Press.

An excellent resource for individuals attempting to patent on their own. It provides examples of documents to be submitted to the patent office as well as examples of related agreements.

Schechter, R.E. and Thomas, J.R. (2003). *Intellectual Property: The Law of Copyrights, Patents and Trademarks*. St. Paul, MN: West Publishing Company.

This text has in-depth coverage of present laws and future issues in intellectual property. In particular, it includes a section on trade secret law that you might refer to when considering patents vs. trade secrets.

Stim, R. (2002). *License Your Invention,* 3rd ed. Berkeley, CA: Nolo Press.

A guide for individuals who are interested in licensing their ideas. It provides guidelines, considerations, and examples of invention licensing.

★ *Key Points*
*Chapter 9*

- Planning is essential for a successful entrepreneurial business.
- Before you decide to move forward with your business idea you should complete a start-up analysis.
- To evaluate the potential of a business and to obtain support you will need a business plan.

# Chapter 9

# PLANNING YOUR BUSINESS*

*Robert J. Loring*

## 9.1 INTRODUCTION

Historically, the process of planning, financing, launching, and success-fully managing a new business venture was a rare and infrequent event. The average new entrepreneur was a male in his late thirties who had worked for corporate America for 10 to 15 years to gain the business experience, exposure to entrepreneurial opportunities, and the business acumen needed to successfully manage a new business venture. The venture capital industry (persons or organizations who fund start-up busi-nesses) was in its infancy in the 1960s, and start-up seed capital funding was extremely difficult to obtain. Table 9.1 shows the growth of the venture capital industry from its inception shortly after the second world war. Almost all new business ventures were initially funded with the entrepreneur's own assets plus those that he could gather from friends and family. This is commonly known as "bootstrapping."

In the 1960s and 1970s the U.S. started fewer than 50,000 new businesses per year. By 1983, this figure had jumped to over 100,000 per year and reached a peak of slightly over 251,000 new business start-ups in 1986. This figure has fluctuated between about 150,000–200,000 new business start-ups per year since that peak period. The relationship between the total amount of venture capital invested and new business start-up activity is not obvious from the table. Since the amount of venture capital has increased dramatically over the past 10 years, but the number of yearly new business start-ups has remained relatively constant, the average amount invested in each new venture capital-backed company has increased significantly over prior levels.

---

*Reprinted with permission from *The Modern Entrepreneurial & Business Planning Process,* Robert J. Loring, Ph.D., 2002, Aztec Press.

| Year | Amount of Venture Capital[a] | Number of New Business Start-Ups[b] |
|------|------------------------------|-------------------------------------|
| 1946 | 0 | |
| 1960 | $10 million | |
| 1968 | $30 million | |
| 1975 | $10 million | |
| 1980 | $600 million | |
| 1983 | | 100,868 |
| 1984 | | 102,329 |
| 1985 | | 249,770 |
| 1986 | | 251,597 |
| 1987 | $4 billion | 233,710 |
| 1988 | | 199,091 |
| 1989 | | 181,645 |
| 1990 | $2.9 billion | 158,930 |
| 1991 | $2.3 billion | 155,672 |
| 1992 | $3.8 billion | 164,086 |
| 1993 | $4.5 billion | 166,154 |
| 1994 | $3.7 billion | 188,387 |
| 1995 | $5.6 billion | 168,158 |
| 1996 | $11.3 billion | 170,475 |
| 1997 | $14.8 billion | 166,740 |
| 1998 | $19.8 billion | 155,141 |
| 1999 | $54.5 billion | 151,016 |
| 2000 | $102.3 billion | |
| 2001 | $37.6 billion | |
| 2002 | $25 billion (projected) | |

**TABLE 9.1**
History of Venture Capital
Investments Relative to
New Business Start-Ups in
the U.S.

[a] Source: National Venture Capital Association (www.nvca.com), *Wall Street Journal* (10/1/02).
[b] Source: The Dun & Bradstreet Corporation: *Business Starts Record* and *A Decade of Business Starts.*

Moreover, women have emerged as the preeminent entrepreneurs in our economy, accounting for more than 60% of the yearly new business start-ups. Without a doubt, the advent and explosion of the Internet, the World Wide Web, and information technology tools have sharply reduced both initial start-up funding costs for new business ventures and ongoing operating costs. The productivity and efficiency gains that information technology and the Internet have afforded the world are irreversible and will become embedded into the fabric of almost every new business that is created henceforth. Today, many new businesses are launched with modest resources, even from the home office environment with just a personal computer, a modem, and an Internet account. That is a truly profound development that was not possible just 10 years ago. However, it is important to bear in mind that the ease of finding significant early stage investment capital for an entrepreneurial venture is very sensitive to the general health of the economy. This fact is certainly evident with just a quick glance at the figure showing the precipitous drop in venture capital funds from 2000 to 2001, reflecting the recession the U.S. economy experienced during that period.

However, even in the face of the largest and longest economic expansion this country has seen in over 50 years, there have been a few negative trends and disturbing events. First, although the U.S. is starting over 150,000 new businesses per year, over 80% fail by the fifth year in business, regardless of the amount or source of financing obtained. Additionally, personal and business bankruptcies are at an all-time high. In 2000, there were 1,253,444 bankruptcy filings in the U.S., 97% of which were personal, rather than business related. This figure has seen a doubling over just the last five years.

Another phenomenon began to emerge in 1998: the prospect of obtaining initial seed capital financing of a new business venture through the initial public stock offering (IPO) process. During the second half of 1998 through the first half of 2000 there were 912 new businesses that received their initial seed capital financing through IPOs (Commscan Equidesk, n.d.). Hundreds of these firms have since gone out of business or have filed for bankruptcy. The obvious question is why did so many businesses that were unrelated to each other, selling different products and services in completely different industries, with different CEOs and management support teams all go out of business during the same relatively short time period? The answer lies in the manner in which these new business ventures were initially financed. The one common thread that all of them shared, from e-Toys to Garden.com, is that these new businesses were prematurely financed through the IPO process from essentially nothing more than a business plan. Very few of them had any long-term history of profitable business operation, a steady flow of revenues, a growing base of customers, or experienced entrepreneurs. This premature form of equity financing removed the pressure and incentive from the CEO to create a profitable and sustainable business model in the normal fashion, that is, slowly over time.

It is important to emphasize that the act of launching any entrepreneurial business venture is a substantial undertaking in terms of time, effort, and funding required to give the venture the highest possible chance for long-term success. Recognize that there are several choices an individual who has an idea or concept for a possible new business venture can make. Each of these choices can be viewed as a fine balance between risk and reward. In the business world there is a direct proportional relationship between the amount of risk an entrepreneur is willing to take and the potential future financial reward that the venture could yield. Our focus throughout this chapter will be on how to start a new business venture yourself. This scenario can take many forms, including a "tech transfer" project, which enables a product developed in a university environment to be transferred to the commercial marketplace. It is unarguably the most exciting approach, and although the most risky, also implies the greatest potential for financial return and other forms of success.

Under normal circumstances, an entrepreneur finances a new business venture with limited equity capital raised from the entrepreneur's own resources. He or she then takes the time to prove the basic business concept and builds a profitable business model. The single most important factor that can predict the future success of a new business venture is the personality of the entrepreneur. This factor is much more important to the successful launch of a new business venture than the amount of financing obtained or the essence of the new business idea or concept or the conditions of the marketplace that the entrepreneur will be competing in. Once the right fit or project is found, the process of planning the launch of the entrepreneurial business can start in earnest.

## 9.2   AM I AN ENTREPRENEUR?

A common entrepreneurial myth is that all entrepreneurs are born with a fixed set of entrepreneurial genes that predisposes them to start new business ventures. In reality, there is no preset entrepreneurial mold that successful entrepreneurs conform to. Rather, a combination of factors produces successful entrepreneurs. Some of these factors do appear to be genetic in origin (for example "likes risk-taking behavior"); others are purely environmental. Many of us can recall someone from our childhood who may have possessed many of these qualities and subsequently went on the start a successful new business venture. You can administer the entrepreneurial test on yourself by asking which of these factors describes your personality best.

### Common Personality Characteristics of Successful Entrepreneurs

| | |
|---|---|
| Self-motivation | Likes risk-taking behavior |
| Meticulous attention to detail | Salesmanship ability |
| Ability to create the illusion of stability | Leadership ability |
| Excellent communicator | Tolerance of ambiguity |
| High market knowledge | Good negotiation skills |

This is by no means an exhaustive list; you can probably think of other personality traits that successful entrepreneurs share. However, out of these ten traits one stands alone as the most vital that every successful entrepreneur must possess; it is "likes risk-taking behavior." This trait cannot be learned or taught; it is one of the few characteristics that you are either born with or not. Many successful entrepreneurs would even go one step beyond this definition and say that they are addicted to risk. In the end it is this common thread of assuming sometimes very significant amounts of risk, both financial, personal, and social, that ties all successful entrepreneurs together.

## 9.3   PLANNING AN ENTREPRENEURIAL BUSINESS

In today's global marketplace, it is now more important than ever before to have a well-thought-out and properly documented written business plan in place prior to launching an entrepreneurial business venture. It is not uncommon, at the conclusion of the business planning process, to decide to pass on the original idea or concept and focus on another idea to develop commercially. Therefore, your first objective in preparing a formal business plan from an original idea is to decide if it is worthwhile to launch the venture at all. Issues such as marketplace conditions, amount of financing required, and revenue and profit projections will all weigh in as ultimate deciding factors.

As you begin the process of formal business plan preparation, keep in mind that through the finished business plan you should achieve the following four main objectives:

- To develop and clearly document the business' competitive advantages when compared to the competition and the current methods
- To accurately estimate the business's future financial performance and determine the amount of financing capital needed
- To use the business plan as a vehicle for raising outside financing capital
- To develop a list of short-term and long-term business goals and objectives to be achieved over the first 5 years of operation of the business

## 9.3.1 Start-Up Analysis

The process of researching and preparing a professional-quality, investment-grade business plan is a comprehensive undertaking, usually consuming 3 to 6 months of an entrepreneur's time. Prior to making this investment in time, effort, and cost, it is wise to perform a relatively quick start-up analysis of the idea or concept. This short process should take approximately a week. Your goal is to assess the five critical factors that will impact your new business idea:

- **Gross Profitability:** At this earliest of stage in the planning process, the new business concept, if successful, should have the potential of generating a minimum of a 50% profit margin. You determine this by estimating what you think your unit product sale might be in dollars and comparing this figure to the major costs associated with generating that unit sale. The process of initial cost and market analysis is discussed in Chapter 5. Although the minimum 50% profit margin might seem high at this stage, it is reasonable because once you actually analyze the marketplace and prepare formal financial projections, this number will invariably decrease. If the business plan is then launched as a going concern, the actual profits will decrease again. There are many factors and costs that simply cannot be accurately estimated within the framework of a business plan that will have an impact upon your new entrepreneurial business post launch. Your goal is to program in enough of a profit buffer so that the realities of the marketplace do not take too large a toll on the new business venture.

- **Industry Analysis:** Ask yourself, "What overall industry will my company's products be competing in, how big is this industry, and what are the current trends and dynamics within the industry?" For example, if you are developing a new bioengineered drug for cancer treatment, your product would fall within the U.S. health care industry, which had over $1.3 trillion in activity in 2000 and is currently growing at a rate of about 7 to 8% per year. The goal is to situate your product within an industry that is as large as possible and that is in a state of growth or expansion, rather than one that is either relatively small or exhibiting no growth. The former scenario will make it much easier for your entrepreneurial firm to get marketplace traction, successfully compete, and exhibit more future potential growth within the industry.

- **Competitor Analysis:** Ask yourself, "What other firms are selling the same or similar products or services as my company's products?" Is this a highly fragmented marketplace with hundreds or even thousands of competitors that each maintain only a fraction of the target market? Or is the competitor environment characterized by a relatively few large and dominant firms with vast corporate assets and resources? It is almost always preferable for a new entrepreneur to enter a marketplace that is fragmented and unstructured, with many competitors. This type of market is more receptive to a niche product or service and yields more market opportunity than one with major dominant firms.

- **Personal Resources:** Almost all new entrepreneurial businesses are initially financed through the entrepreneur's own resources and assets, possibly with the help of friends and family. Ensure that you will have access to the minimal amount of start-up capital needed to formally establish your business and begin the process of marketing your company's products or services. This initial amount of "pre-seed" capital is usually a relatively low number, perhaps $10,000 to 25,000, and will fund the initial start-up of the firm for a few months. If you will not have access to this amount of basic financing it is probably not worthwhile investing 6 months of your time preparing the formal business plan.

- **Technical Feasibility:** Ask yourself, "Is my business concept or idea predicated on some future technology that has not yet been invented?" If so, you should definitely pass on the idea and wait for the needed technology to develop. You could do some early-stage market research about the business concept, but not much beyond that, since any long-term financial projections, requirements, and organizational strategy will most likely be affected by the technology that you are waiting for. You would have great difficulty obtaining outside financing of any type for a business plan that is based on a nonexistent technology.

### 9.3.2 The Formal Business Planning Process

Once you have completed the start-up analysis of your new business idea as detailed above and you have a clear understanding of the goals and objectives of the formal written business plan, you are ready to begin the serious analytical work and preparation process. The good news here is that there is a standard business plan outline or template to follow for preparing a business plan regardless of the nature of the business idea or concept. All business plans should follow this standard format unless there are compelling reasons to deviate from it. This format includes six formal sections: an executive summary, a detailed product description, a comprehensive market analysis and marketing plan, a coherent set of financial projections and financing requirements, an organizational strategy and management section, and, a set of appendices. The end result should be a well-thought-out, well-researched, and professionally prepared document that is about 25 to 40 pages in length, excluding appendices. Many new entrepreneurs make the mistake of thinking that if they prepare a 150-page-long business plan it will somehow give them an advantage in competing for financing capital. Unfortunately, the exact opposite is true. To understand this apparently contradictory phenomenon, you need to be able to put yourself into the mind of a prospective investor in your business. If you submit a 150-page business plan that weighs 4 pounds, the first thing that a prospective investor is going to think is, "What is it about this business concept that is so complicated or convoluted that the entrepreneur cannot explain it clearly and concisely in the standard 25 to 40 pages?" Furthermore, if it really does take 150 pages to explain the business idea, an investor may conclude that few people are going to be able to understand it, especially the company's likely customers. This is obviously an unfavorable impression to make. The second question that will go through an investor's mind is, "What is it about the entrepreneur himself or herself that this person has trouble communicating the essentials of the new business idea within the standard prescribed length?" If the investor feels that the entrepreneur is a poor or lackluster communicator, this could spell trouble on many future fronts should the business actually be financed and launched.

This will be the shortest section of your business plan. It should be no more than one or two pages in length. It is to be written last but placed first in your final written business plan. Its sole purpose is to highlight all of the major points and salient features of the business plan and above all to entice the prospective investor to read the entire business plan. Do not make the mistake of trying to compose the executive summary first over the course of a weekend in order to get it out of the way. This will result in the executive summary not tying in well and supporting the detailed analysis that follows. This undesirable result will be patently clear to any experienced reviewer of your business plan.

The executive summary should include the following categories:

- **Product Description:** Briefly describe the proposed new product. This should be a cursory one- or two-paragraph description of the business product and proposed business model with enough detail that the reader will understand the basic concept or product to be sold. See the discussion of the abstract in Chapter 6.
- **Competitive Advantages:** It is crucial for your new business to have discernable and definable competitive advantages when compared to the current methods or competitors' products. If not, your new business may be very short lived. Describe these differentiating factors early in the business plan and reinforce them frequently.
- **Your Industry and Target Market:** Briefly describe the overall industry your entrepreneurial business will be competing in, its size, and current growth dynamics. Describe the specific target customer market within this overall industry in terms of its estimated size, your anticipated customer categories, and any other important demographic factors.
- **Financial Projections:** In summary fashion, clearly state the projected financial performance and returns over the first 5 years of successful operation of your business. State the expected revenues, profits, and cash flow. Indicate the total amount of financing required to achieve these projected returns and what you feel the best form of the financing should be.
- **Exit Strategy:** If your entrepreneurial business will require outside financing from the venture capital or angel investor community, then a clearly articulated exit strategy is essential. This is the plan for how the investors will be repaid for their investments. This could take the form of an IPO, an asset or stock sale, or a liquidation of the company's assets.
- **Management Strength:** Describe the relative strengths of the management team in terms of prior start-up new business and technical experience and any other important factors.

## Product Description

The objective of the product description section of your business plan is to provide the reviewer with a detailed description of your entrepreneurial business' proposed products or services, so that a clear understanding exists of exactly how the business will be operating. This is parallel to the project proposal described in Chapter 6. You may use much of the background and motivating material from your proposal and reports. Describe both the product and how the company will produce the product, sell it to your anticipated customers, and generate resulting profits and positive cash flow. If your company will require a key strategic partner, subcontractor, or outsource provider, that relationship must be described fully in terms of the nature of the proposed relationship and the projected incurred costs.

If your company is producing a physical product, use the list of product features below as a guideline in preparing a cogent product description. Many of these features will be important to a product company. Your challenge is to look over the list and develop a strategy around those items that are relevant. Those that are not directly relevant to the proposed product's success can be ignored. These factors often serve to begin the process of specifying the business's competitive advantages. Sample quality dimensions of products are:

- **Performance:** operating characteristics
- **Features:** important special characteristics
- **Flexibility:** meeting operating specs over some time period
- **Durability:** amount of use before performance deteriorates
- **Conformance:** match with pre-established standards
- **Serviceability:** ease and speed of repair or normal service
- **Aesthetics:** how a product looks and feels
- **Perceived Quality:** subjective assessment of characteristics

In a similar fashion, the list below includes features that tend to be important success factors for service-related businesses. Again, your challenge is to examine this listing and build a strategy around achieving those performance features that will influence purchasing decisions by your anticipated customers. Sample quality dimensions of services are:

- **Timeliness:** performed in promised time period
- **Courtesy:** performed cheerfully
- **Consistency:** all customers have a similar experience each time
- **Convenience:** accessible to customers
- **Completeness:** fully serviced, as required
- **Accuracy:** performed correctly each time

As you prepare the product description section you can use the checklist below of important questions to answer:

- Who are your customers?
- What need or niche is the product or service meeting?
- If the product is to be manufactured, how will that be accomplished?
- If the service requires a major subcontractor, how will that be achieved?
- How will the product or service reach your market niche?
- How does the proposed product compare to current methods?

It is often helpful to walk the reader completely through how a sample transaction or unit sale would be processed by your new company, using before and after examples wherever possible. This approach is especially useful for businesses producing an intangible product, such as information technology, Internet, or e-commerce ventures where the product might be some type of back office, middleware, or infrastructure support service, rather than a tangible product. Use of flow charts, graphs, diagrams, and tables is often very helpful in explaining how the business model will actually operate. You should thoroughly describe the current methods or competitive products that exist to meet the current market demand. This will naturally dovetail into a comparison of product features between competitors and an opportunity to again underscore your company's proposed competitive advantages.

## Market Analysis

The first concept that needs to be understood when beginning the process of collecting market analysis data, and developing subsequent marketing plans, is that the terms *market analysis* and *marketing plan* are not equivalent or interchangeable. As we will see, performing a thorough market analysis is the first step in the process. You will invest more time preparing the market analysis than any other section of the business plan. It is also common that this part of the formal business planning process represents the first time the entrepreneur might get the impression that the commercial viability of the original idea or concept is not as strong as first estimated.

A market analysis is the collection of data about a particular industry or marketplace for which the entrepreneur is considering developing a competitive product. As you collect the plethora of data that you will analyze and subsequently use, you should keep one end point in mind: What you are trying to do in the market analysis is to simply define who the customers are for your proposed product or service. The answer to this question will form the basis for the marketing plan to follow and serve as the starting point for developing all of the financial projections for your business plan.

It is frequently helpful to think about this process of defining the customers by envisioning the marketplace as a set of concentric circles, with your specific target market at the very center of the set of concentric circles. Your goal, therefore, is to distill down the large overall industry into the well-defined target market of customers to whom you will be selling your entrepreneurial product.

The primary data collection tool used in the modern market analysis process is the Internet and World Wide Web. With over 200 million different Web sites in existence there is an abundance of information available to entrepreneurs planning new business ventures. Your first task is to define the largest concentric circle — the overall industry in which you are planning to situate your entrepreneurial business. The larger the overall industry, the more upside potential the business venture will tend to have. Hence it is desirable to try to pick an industry that not only is relatively large, but also one that is growing. Remember, growing markets are very forgiving of mistakes or missteps that you may make at the outset of a new venture. Here are some examples of different industries and their sizes:

- Information technology: approximately $400 billion per year
- U.S. health care industry: approximately $1.8 trillion per year
- U.S. pharmaceutical industry: approximately $150 billion per year
- U.S. Defense Department expenditures: approximately $274 billion per year

Sometimes the size and characteristics of the overall industry are not readily apparent. For example, if you were trying to develop a Web-based software solution for expediting and reducing the costs of warehousing, order fulfillment, shipping, and collection that could potentially be used by any consumer product company, the exact industry being served would not be obvious. The product, by design, could be used by almost any consumer product business. In a case like this other metrics and indices need to be considered, such as the size of competitors' businesses and the specific resources you are able to garner to substantiate the financial return projections you make.

The second step in the market analysis process is to accurately define your company's target market. The target market for any company's products represents the specific subset of prospective customers from the overall industry that the new product will be directed toward. No company markets its products to an entire industry. The target market is typically a small subset of the overall industry, perhaps up to 15% in terms of dollars or units sold. This is a very important determination to make, as it will influence both the subsequent marketing plan and the financial projections to follow. You can use whatever factors you feel are important in segmenting the overall industry into a target market for your company's products.

> An example of one target market can be found in the mountain biking industry. The worldwide market for mountain bikes is approximately $5 billion dollars. Within this industry there is a distinct segment of premium mountain bikes that sell for over $3000 each, representing about 15% of the overall market or about $750 million in annual sales. If you were developing an entrepreneurial product, say an all-wheel drive mountain bike that might sell for $3000, you would want to focus your marketing efforts and financial projections on just this segment of the overall industry. The $750 million segment is your target market. In this example, a price demographic was used to segment the overall industry.

To start the process, simply ask yourself who your customer categories are for your new product. The good news here is there are only four:

- Individuals who purchase consumer products
- Corporations that purchase business products or services
- Educational institutions, such as schools and universities
- Government agencies, such as the Department of Defense or Veterans Administration

These customer categories are not mutually exclusive; it is possible to target more than one of them, but only choose those that are relevant for your entrepreneurial product. You can further break down and subdivide these customer categories according to demographic factors that will be important for your specific entrepreneurial product. A demographic factor is any factor that will help you subdivide the industry into a smaller segment containing just those specific customers for your entrepreneurial product or service. The better defined your target market is, the more focused will be your marketing and advertising plans and resulting financial projections. For consumer products, this could include gender, age, income level, location, and education. For corporate, educational, or government customers, demographic factors could include type of business, location, and size.

Once you have defined these potential customer groups and subgroups, your task is to find out as much about their preferences as possible. This process known as "target market characterization" is accomplished by entrepreneurs and established firms introducing new products, in four different ways: test marketing, focus groups, interviews, and surveys. New-product test marketing and focus groups are tried and proven long-term and expensive methods for collecting market analysis data prior to the formal launch of a new product. Test marketing involves selecting a small demographic or geographic sampling of your proposed target market and test launching the new product to this selected small group. Sometimes this is done simultaneously in several demographic or geographic locations. One of the most recognizable industries that frequently employs test marketing is the fast food industry. A test marketing session may be scheduled for as little as a month to more than a year, depending on the nature and extent of customer feedback data needed.

Focus group market data collection is frequently used by consumer product companies, such as the toy and cosmetics industry, when considering the national or international launch of a product. With this method, simultaneous focus groups of perhaps 20 or more potential target-market customers are assembled. The company representatives then fully explain and demonstrate the new product and ask for candid feedback from the audience. The data collected is culled, processed, and analyzed, helping the entrepreneur make adjustments to the product prior to formal launch. Because of the long-term and costly nature of both focus groups and test marketing, these methods are not appropriate for senior design, although you could indicate that one of these methods might be helpful in defining the potential market should the business plan receive initial seed capital financing.

Alternatively, market surveys and interviews are viable methods for collecting market data for prospective customers in the context of senior design and industry projects. Market surveys are done in written form and can be distributed either in person or through e-mail. The idea is to prepare a brief written questionnaire as described in Chapter 5 that contains no more than five or six important questions you would like feedback on from actual prospective customers. The questions should be based on the perceived competitive advantages and important product features of your new entrepreneurial product that will help to differentiate it in the competitive marketplace. You should try to anticipate exactly what will drive the purchasing decisions for your entrepreneurial product. In general, a prospective customer will have one of three needs in mind when making a purchasing decision. These are unique product features such as those offered by Microsoft or Federal Express, low prices such as those offered by McDonalds or the U.S. Postal Service, or perhaps a combined product feature–low price strategy like that of United Parcel Service or Dell Computer. As shown in the example in Chapter 5, the technical questions should be aimed at getting answers about these important issues.

The last question in the survey should be financial in nature. You should plan to ask a direct question about the amount of money you are planning to charge for the product and whether this preliminary number would be acceptable to your potential customers. This does not mean to get out the dart board, close your eyes, throw a dart, and see where it lands in order to determine the best initial product pricing. Rather, there is a tested and proven method for making this initial determination of pricing strategy.

As a first step you need to consider exactly what pricing information you are trying to determine from the marketplace. This optimal number is called the "minimax," and it represents the highest price you can charge for your entrepreneurial product or service, while only losing the minimal number of prospective new customers. This number is arrived at by conducting a subgroup analysis and plotting the resulting data on a graph. Your goal is to create perhaps four or five subgroups from the original list of prospective customers for your survey based on different price points for your prospective product. To determine these different prospective price points for the survey, you must look at two factors:

- First determine what the minimum costs will be to produce your entrepreneurial product, then add to that number factors such as 50%, 100%, 150%, and 200%. These new numbers will be the specific prospective prices that you will ask of those who complete your survey. For example, if an entrepreneur has developed an ocular implant for curing presbyopia at an estimated unit cost of $100 per one thousand units, the end users surveyed would be asked, "Would you be willing to pay $150, $200, $300, or $400 for this product exactly as represented?"
- The second factor is to survey the competition's pricing for the same or similar products and use this information to help adjust or guide you as you prepare the subgroup pricing questions.

As the survey results are collected and tabulated, a definite trend will emerge that will guide you as to best initial price for your entrepreneurial product. A sample hypothetical subgroup analysis is shown in the example below. For this example, four groups of 20 people each were asked if they would pay $8, $12, $16, and $20, respectively, for a new entrepreneurial product. At $8, 19 people, or 95%, indicated acceptance of the product. At $12, there was a slight drop-off to 90% acceptance. At $16, 75% of those surveyed would buy the product, and finally at $20, only 55% would purchase. The conclusion is that the new product should probably be initially priced somewhere between $12 and $16, but definitely not at either $8 or $20. Although an $8 price would capture 95% of the market, too much money would be "left on the table" unnecessarily. A $20 price would result in the highest unit profit margin, but at the expense of losing 45% of the potential new market. Neither of these two extremes is acceptable for a new venture. The real emphasis here is to provide a good pricing guideline so that you can intelligently select the best initial price and not deviate either too high or too low. Bear in mind that as soon as the product is actually introduced, the marketplace will react, including both a customer and competitor reaction. This often results in subsequent pricing adjustments.

─────────────────────────────▼─────────────────────────────

## SAMPLE SUBGROUP ANALYSIS TO DETERMINE MINIMAX

Acceptance of a proposed product based on interview/survey of four groups of 20 people each

| Price | Group 1 | Group 2 | Group 3 | Group 4 |
|-------|---------|---------|---------|---------|
| $8 | 95% | | | |
| $12 | | 90% | | |
| $16 | | | 75% | |
| $20 | | | | 55% |

─────────────────────────────▲─────────────────────────────

The final task in the market analysis process is to analyze and report on the competitive products that currently exist. The first issue to understand is that there is always competition. Sometimes, the competition is overt, meaning you are able to locate and profile other companies that provide the same or similar products as your new company. Other times the competition is covert, meaning there are apparently no direct competitors for the new venture's proposed products. When this latter scenario emerges, you have to ask yourself,

"Where are my prospective customers spending the money that I will ask them to spend on my product?" The answer to that question determines your competition. It may be a totally unrelated discretionary product or service, but nonetheless, it still represents competition for your customers' dollars.

The Internet can be used as the primary competitor research tool. Today, a great many companies have a Web presence, whether a simple digital brochure or a structured Web site conducting e-commerce sales of the company's products. Competitors that are publicly traded companies will have a plethora of information, both financial and technical, available on their Web sites. Your job is to profile in detail perhaps four or five direct competitors for your entrepreneurial product, both in narrative form and with a competitive matrix. A sample competitive matrix is shown below. The competitive matrix is designed to contrast how your proposed product will compare to four or five competitive products according to factors that you project will give your product competitive advantages over the competition and subsequently drive product sales to your firm.

As in the example, list your company's name and the specific names of competitor firms that you have located. Next, list five or six factors that will favorably compare your company to the known competition. These factors will differ from one enterprise to another and again should represent what the company's competitive marketplace advantages will be when the product is launched. Once the table is set up, you need to honestly rate your company and the competitor firms on a simple scale of one to five, with one being the lowest and five being the highest rating for that factor. Once the scores are assigned, total them for each company profiled. The total score for your company should be clearly superior to that of the competitors. There is no absolute requirement here, but you should seek to make a positive visual. Often, the preparation of the competitive matrix is the first time during the formal business planning process that the entrepreneur gets a clue that the original idea or concept might not be as commercially viable as anticipated. This would be the case if the total scores differed by only a few percentage points. We call this condition "competitive parity." Should this be the result, you will want to reexamine those factors that you chose; perhaps an important competitive advantage was not properly presented or did not surface in the planning process.

## SAMPLE COMPETITIVE MATRIX

| Factor | Your Company | Competitor or Technology | | | |
| --- | --- | --- | --- | --- | --- |
| | | A | B | C | D |
| Low price | 5 | 3 | 3 | 3 | 2 |
| Superior quality | 5 | 4 | 4 | 4 | 3 |
| Customizable | 5 | 3 | 3 | 2 | 3 |
| Unique features | 5 | 5 | 4 | 4 | 1 |
| Rapid product delivery | 4 | 3 | 3 | 2 | 3 |
| **Total** | **24** | **18** | **17** | **15** | **12** |

The final question you must answer before concluding the market analysis is, "What percent of the target market can my entrepreneurial business capture at the end of 5 years of operation, assuming that the needed financing is secured, the right people are hired, and the infrastructure is built-out as forecasted?"

Before making this assessment, you need to look over all of the market analysis data you have gathered about the industry, the marketplace, the prospective customers, and the competition. The dollar figure you arrive at should not exceed 15% of the target market size in terms of either dollars or units sold. You are free to adjust this figure based on the number of projected units to be sold and the resources you will be able to garner for the effort. The reason for this guideline is that although it is possible for a very successful new product to capture perhaps 40 or 50% of a target market after 5 years of highly successful operation, it is not reasonable to make this assumption when preparing a business plan under any but the most unusual circumstances. A prospective reviewer or investor for your business would conclude that you are not in touch with the marketplace or the realities of managing a successful new business venture. Clearly, you do not want to cast this kind of impression. The final dollar amount will represent your company's fifth-year projected revenues and will be the starting point for the build-out of the financial proforma to follow.

### Marketing Plans

As with conducting a market analysis, the key objective is to be able to define exactly who will be the customers for your new entrepreneurial product. Development of a marketing and promotional plan has a similar key objective: to establish the best methods for the distribution of the new product to those customers you have defined and characterized in the market analysis. Your objective will be to use and leverage those competitive advantages that surfaced in the market analysis as the basis for formulation of a marketing and promotional strategy.

This process begins with an accurate description of the proposed product, now focused on advertising to attract customers. The unique product features that were demonstrated as a result of the market analysis should be fully utilized in preparation of marketing and promotional materials. Be sure to describe any significant advantages of your proposed product compared to the competitors' products you profiled in the competitive matrix. Your marketing plan must also incorporate the initial pricing strategy for your entrepreneurial product. Obviously, the initial price for your new product should be the minimax price you determined from the subgroup analysis survey data collected.

Next you will need to decide on a distribution strategy for your new product. This is how you will fill customer orders, distribute your product, and provide aftermarket service and other customer service–related matters. Your distribution choices include the Internet, as well as conventional approaches such as brick and mortar establishments, direct product sales, wholesaling, and licensing. In general, use the Internet wherever possible. The Internet represents the most cost-effective and efficient approach for distribution of any company's product, so use it liberally throughout your company's supply and value chain.

Your final task in developing an overall marketing plan is to devise a promotional and advertising plan designed to both reach your target-market customers and convince them to buy your product. The advertising plan should be focused squarely on those target-market customers' needs that you have characterized and defined as part of the market analysis, specifically incorporating the key factors that will drive purchasing decisions by your target market customers. Again, this will take the form of either unique product features, cost leadership, or a combined cost leadership–unique features model.

You should plan to develop both a conventional and virtual advertising plan. Below is a list of common forms of both conventional and virtual advertising methods. Your task is to select the most appropriate methods from both lists and build a promotional strategy around them. Direct mail is an effective form of advertising but is one of the most expensive forms as well. A well-designed direct mail campaign might produce up to a 2% response rate. This would be viewed as a huge success in light of the thousands of direct mail pieces that would be sent. In general, if your customer base is nonconsumer (business, government, educational) you should seriously consider incorporating direct mail into your advertising strategy. There will be substantially fewer corporate, educational, or government buyers compared to the consumer market, and the higher cost of direct mail would be justified for these categories of customers. Some virtual and conventional advertising methods are given below.

## Virtual Advertising Methods

- Banner advertising
- "Pay per click" advertising such as Google and Overture
- Search engines
- Company public Web site
- Online government bid lists
- Formation of strategic marketing alliances, that is, cobranding

## Conventional Advertising Methods

- TV
- Radio
- Print
- Trade shows
- Direct mail
- Business cards
- Brochures
- Yellow pages

## Preparing Financial Projections and Requirements

When preparing a financial proforma for your new entrepreneurial business you must determine:

- How much investment capital will this new business require to get it to the point of positive cash flow or profitability?
- What is the best form of financing (equity, loans, grants)?
- What is the best source or source for financing (venture capital, angel investors, commercial banks, government granting agencies, or family and friends)?

In order to answer these critical questions, we must be able to estimate the new business' anticipated capital equipment requirements, projected revenues, projected expenses, projected profits, and projected cash flows. The tool that will yield this financial data is a 5-year financial proforma. A proforma is a statement of future projected sales, expenses, profits, and cash flows. Five years is the standard yardstick that is used for almost all new entrepreneurial products, because it represents the period of highest attrition for new business start-ups. If your business is able to survive the first 5 years, you are probably going to have a viable business for many years ahead, barring such problems as unpredicted major changes in the marketplace or operating costs as a result of government regulations. As you are preparing and building out this proforma, keep in mind what it is likely to be used for: a tool to secure outside financing, whether in the form of equity capital, loans, or grants.

As we discuss the process, you should refer to the sample financial proforma in the following examples. Before formally preparing the financial proforma, you must first estimate the expected capital equipment costs (CEC) for your entrepreneurial business. Capital equipment includes items such as computers, buildings, cars, specialized processing or production equipment, and furniture. On a separate document, list these items along with their estimated costs and generate a total CEC. Keep this figure; you will need to refer to it as you prepare the financial proforma.

### Preparing the Financial Proforma

Often a new entrepreneur approaches this part of the business plan by trying to somehow estimate what the costs and expenses will be to launch the business and run it, without first estimating the expected sales. This would be a serious error, because the only factor that drives any company's expenses are its sales. If there are no sales, there will be no expenses. Therefore, the first step is to estimate your new entrepreneurial business's projected sales or revenues over the initial 5 years of operation. We start by estimating the fifth-year projected revenues first. This is the same figure you estimated at the conclusion of the market analysis. It has now become the starting point for developing the 5-year financial proforma. To reiterate, this critical estimate should not represent more than a 15% share of your target market, in terms of both revenues and units sold. The resulting fifth-year sales, profits, and cash flows will form the basis of determining your new company's eventual market value, which will directly influence the amount of outside financing you are able to secure for the business.

To get started, consider a hypothetical entrepreneurial business with an overall market size of $1 billion, a target market size of $100 million, and a revenue estimate of 15%, or $15 million, after 5 years of successful operation. This $15 million revenue estimate is now entered directly onto the financial proforma spreadsheet as the fifth-year revenue projection. As you examine the proforma format, note that the estimating and reporting periods are monthly during the first year, quarterly during years 2 and 3, and yearly during years 4 and 5. This is the convention because it becomes increasingly difficult to make exact financial projections the farther out you look. At this stage you also need to consider where you anticipate the revenues to come from. Recall from the market analysis that you defined certain customer categories, such as consumers, businesses, government agencies, or educational institutions, for your entrepreneurial business. We also examined what form the sales would be to these customers. This included unit product sales, membership subscriptions, service contracts, and licensing revenues. Ask yourself a simple question: Will any of these customer categories or forms of revenue represent more than 20% of my total estimated revenue during the initial 5 years of operation? If the answer is yes, you should break out those categories and forms of revenue from the total and itemize them separately on the revenue projection section of the proforma (Figure 9.1 and Figure 9.2).

**FINANCIAL PROFORMA YEAR ONE**

| | JAN | FEB | MAR | APR | MAY | JUN | JUL | AUG | SEP | OCT | NOV | DEC | TOT YR1 |
|---|---|---|---|---|---|---|---|---|---|---|---|---|---|
| **REVENUES:** | | | | | | | | | | | | | |
| Business Customers: Contracts | | | | | | | 10,000 | 10,000 | 10,000 | 10,000 | 10,000 | 10,000 | 60,000 |
| Consumer Customers: Product Sales | | | | | | | 52,500 | 52,500 | 52,500 | 52,500 | 52,500 | 52,500 | 315,000 |
| **Total Revenues:** | 0 | 0 | 0 | 0 | 0 | 0 | 62,500 | 62,500 | 62,500 | 62,500 | 62,500 | 62,500 | 375,000 |
| **Units Sold** | | | | | | | 625 | 625 | 625 | 625 | 625 | 625 | **3,750** |
| **OPERATING EXPENSES:** | | | | | | | | | | | | | |
| Rent (2000 sq. ft. @ $20/sq. ft. per year) | 3,333 | 3,333 | 3,333 | 3,333 | 3,333 | 3,333 | 3,333 | 3,333 | 3,333 | 3,333 | 3,333 | 3,333 | 39,996 |
| Employee Expense @ 20% Gross | 900 | 900 | 900 | 900 | 900 | 900 | 2,000 | 2,000 | 2,000 | 2,000 | 2,000 | 2,000 | 17,400 |
| Benefits @ 15% of Employee Expenses | 135 | 135 | 135 | 135 | 135 | 135 | 300 | 300 | 300 | 300 | 300 | 300 | 2,610 |
| Information Technology expense | 1,000 | 1,000 | 1,000 | 1,000 | 1,000 | 1,000 | 1,000 | 1,000 | 1,000 | 1,000 | 1,000 | 1,000 | 12,000 |
| Consulting Expense | 2,000 | 2,000 | 2,000 | 2,000 | 2,000 | 2,000 | 2,000 | 2,000 | 2,000 | 2,000 | 2,000 | 2,000 | 24,000 |
| Web Site | 2,500 | 2,500 | 2,500 | 2,500 | 2,500 | 2,500 | 2,500 | 2,500 | 2,500 | 2,500 | 2,500 | 2,500 | 30,000 |
| Shipping | 500 | 500 | 500 | 500 | 500 | 500 | 500 | 500 | 500 | 500 | 500 | 500 | 6,000 |
| Office Furnishings | 500 | 500 | 500 | 500 | 500 | 500 | 500 | 500 | 500 | 500 | 500 | 500 | 6,000 |
| Office Supplies | 500 | 500 | 500 | 500 | 500 | 500 | 500 | 500 | 500 | 500 | 500 | 500 | 6,000 |
| Advertising | 10,000 | 10,000 | 10,000 | 10,000 | 10,000 | 10,000 | 10,000 | 10,000 | 10,000 | 10,000 | 10,000 | 10,000 | 120,000 |
| Legal/Accounting | 125 | 125 | 125 | 125 | 125 | 125 | 125 | 125 | 125 | 125 | 125 | 125 | 1,500 |
| Insurance | 100 | 100 | 100 | 100 | 100 | 100 | 100 | 100 | 100 | 100 | 100 | 100 | 1,200 |
| Telephone | | | | | | | | | | | | | |
| Interest | 300 | 300 | 300 | 300 | 300 | 300 | 3,000 | 3,000 | 3,000 | 3,000 | 3,000 | 3,000 | 19,800 |
| Travel and Entertainment | 500 | 500 | 500 | 500 | 500 | 500 | 500 | 500 | 500 | 500 | 500 | 500 | 6,000 |
| Miscellaneous | 2,239 | 2,239 | 2,239 | 2,239 | 2,239 | 2,239 | 2,636 | 2,636 | 2,636 | 2,636 | 2,636 | 2,636 | 29,251 |
| **Total Op. Expenses:** | 24,632 | 24,632 | 24,632 | 24,632 | 24,632 | 24,632 | 28,994 | 28,994 | 28,994 | 28,994 | 28,994 | 28,994 | 321,757 |
| **PROFIT/LOSS** | (24,632) | (24,632) | (24,632) | (24,632) | (24,632) | (24,632) | 33,506 | 33,506 | 33,506 | 33,506 | 33,506 | 33,506 | 53,243 |
| **CUMULATIVE GAIN (LOSS)** | (24,632) | (49,265) | (73,897) | (98,529) | (123,162) | **(147,794)** MCL | (114,288) TAP | (80,781) | (47,275) | (13,769) | 19,737 BEP | 53,243 | 53,243 |

**Operating Cash Need (MCL X 1.5):** 221,691
**TAP: Year One, Quarter Three**
**BEP: Year One, Quarter Four**

**FIGURE 9.1**
Sample financial proforma: Year 1.

## FINANCIAL PROFORMA YEAR TWO-FIVE

| | Year 2Q1 | Year 2Q2 | Year 2Q3 | Year 2Q4 | Year 2Tot | Year 3Q1 | Year 3Q2 | Year 3Q3 | Year 3Q4 | Year 3Tot | Year 4Tot | Year 5Tot | |
|---|---|---|---|---|---|---|---|---|---|---|---|---|---|
| **REVENUES:** | | | | | | | | | | | | | |
| Business Customers: Contracts | 75,000 | 150,000 | 125,000 | 100,000 | 450,000 | 175,000 | 250,000 | 125,000 | 300,000 | 850,000 | 3,500,000 | 5,000,000 | |
| Consumer Customers: Product Sales | 300,000 | 600,000 | 1,000,000 | 1,400,000 | 3,300,000 | 1,700,000 | 2,000,000 | 2,500,000 | 2,700,000 | 7,900,000 | 10,000,000 | 10,000,000 | |
| **Total Revenues:** | 375,000 | 750,000 | 1,125,000 | 1,500,000 | 3,750,000 | 1,875,000 | 2,250,000 | 2,625,000 | 3,000,000 | 8,750,000 | 13,500,000 | 15,000,000 | |
| **Units Sold** | 3,750 | 7,500 | 11,250 | 15,000 | **37,500** | 18,750 | 22,500 | 26,250 | 30,000 | **97,500** | **135,000** | **150,000** | **423,750** |
| | | | | | | | | | | Total Proj, Units Sold: | | | |
| **OPERATING EXPENSES:** | | | | | | | | | | | | | |
| Rent (2000 sq. ft. @ $20/sq. ft. per year) | 9,999 | 9,999 | 9,999 | 9,999 | 39,996 | 18,000 | 18,000 | 18,000 | 18,000 | 72,000 | 72,000 | 72,000 | |
| Employee Expense @ 20% of Gross | 15,000 | 30,000 | 25,000 | 20,000 | 90,000 | 35,000 | 50,000 | 25,000 | 60,000 | 170,000 | 700,000 | 1,000,000 | |
| Benefits @ 15% of Employee Expense | 2,250 | 4,500 | 3,750 | 3,000 | 13,500 | 5,250 | 7,500 | 3,750 | 9,000 | 25,500 | 105,000 | 150,000 | |
| Information Technology Expense | 2,000 | 2,000 | 2,000 | 2,000 | 8,000 | 1,800 | 3,600 | 4,200 | 6,000 | 15,600 | 29,400 | 55,200 | |
| Consulting Expense | 12,000 | 24,000 | 36,000 | 48,000 | 120,000 | 72,000 | 86,400 | 100,000 | 125,000 | 383,400 | 694,800 | 1,303,200 | |
| Web Site | 5,000 | 5,000 | 5,000 | 5,000 | 20,000 | 10,000 | 10,000 | 10,000 | 10,000 | 40,000 | 70,000 | 130,000 | |
| Copier | 1,000 | 1,000 | 1,000 | 1,000 | 4,000 | 900 | 1,200 | 1,600 | 2,000 | 5,700 | 10,500 | 19,800 | |
| Office Furnishings | 900 | 1,800 | 2,700 | 3,600 | 9,000 | 5,400 | 7,500 | 9,000 | 12,000 | 33,900 | 62,400 | 117,300 | |
| Office Supplies | 600 | 1,200 | 1,800 | 2,400 | 6,000 | 3,600 | 5,200 | 7,100 | 11,000 | 26,900 | 50,200 | 95,200 | |
| Advertising | 15,000 | 30,000 | 45,000 | 60,000 | 150,000 | 90,000 | 120,000 | 145,000 | 200,000 | 555,000 | 1,020,000 | 1,920,000 | |
| Legal/Accounting | 750 | 1,500 | 2,200 | 3,000 | 7,450 | 4,500 | 6,000 | 8,000 | 11,000 | 29,500 | 54,500 | 103,000 | |
| Insurance | 600 | 1,200 | 1,800 | 2,400 | 6,000 | 3,600 | 6,500 | 7,900 | 12,000 | 30,000 | 56,400 | 106,300 | |
| Telephone | 6,000 | 12,000 | 15,000 | 20,000 | 53,000 | 15,000 | 15,000 | 15,000 | 15,000 | 60,000 | 105,000 | 195,000 | |
| Interest | 4,500 | 9,000 | 14,000 | 18,000 | 45,500 | 12,000 | 15,000 | 18,000 | 24,000 | 69,000 | 126,000 | 237,000 | |
| Travel and Entertainment | | | | | | | | | | | | | |
| Miscellaneous | 7,560 | 13,320 | 16,525 | 19,840 | 57,245 | 27,705 | 35,190 | 37,255 | 51,500 | 151,650 | 315,620 | 550,400 | |
| **Total Operating Expenses:** | 83,159 | 146,519 | 181,774 | 218,239 | 629,691 | 304,755 | 387,090 | 409,805 | 566,500 | 1,668,150 | 3,471,820 | 6,054,400 | |
| **PROFIT/LOSS** | 291,841 | 603,481 | 943,226 | 1,281,761 | 3,120,309 | 1,570,245 | 1,862,910 | 2,215,195 | 2,433,500 | 7,081,850 | 10,028,180 | 8,945,600 | |
| **CUMULATIVE GAIN (LOSS)** | **345,085** | | | | | | | | | | | | |

**FIGURE 9.2**
Sample financial proforma: Years 2 to 5.

Once you have the fifth-year revenues estimated, your job is to back-fill the revenue projections for the remaining 4 years, starting with year one. Note on the year 1 spreadsheet that there are no revenues included for the first 6 months of this hypothetical business. This initial period of time, after the business is legally organized, but before the business records its first revenues, is called the business ramp-up period. All new businesses have a ramp-up period when expenses are being incurred but there are no sales. This period will vary from business to business. Your job is to decide what ramp-up period is applicable to your entrepreneurial business and fill in zero revenues for that initial period.

The remaining monthly revenue projections for year 1 are calculated by taking 5% of the expected monthly revenues during the fifth year. In our example, the projected monthly revenues in the fifth year of business are $1,250,000. Five percent of that figure is $62,500 per month, which reflects the monthly revenues for the last 6 months of the first year. For each subsequent projected financial period take 10%, 20%, 30%, 40%, 50%, 60%, 70%, 80%, and 90% of the fifth-year monthly revenue projection and record this as your estimated revenues for the reporting period considered. This will complete the build-out of the revenue section of your financial proforma. Once you have made these calculations, you are free to adjust them as necessary to reflect your new business' resources, assets, and any other factors that will influence the revenue projection.

The next step is to estimate the expenses that will have to be incurred in order to produce the revenues you have just estimated. This step involves preparing a chart of accounts. This is a listing of the major expense categories your entrepreneurial business will require to operate each month. Do not include major capital expenditures on the chart of accounts. Capital equipment purchases are depreciated, rather than expensed. Alternatively, if you plan on leasing some capital equipment items, then the monthly lease fees would be listed on the chart of accounts. Refer to the spreadsheet example as your guide. Be sure not to duplicate an item on both the CEC list and the operating expense list. There are several expenses to account for as you prepare the chart of accounts. First, every business has a rent expense, even if you are planning to launch the business out of your house. The IRS has created a home office tax form to use for this purpose (Form 8829).

Next, look over the sample expense listing carefully. Use only items that are relevant for your business. More importantly, if there are any operating expense categories that are not on the sample listing, feel free to add them to the list and estimate a cost for each item. Note that the last operating expense category on the sample listing is "miscellaneous." This represents 10% of the sum of the operating expenses for that period. This additional amount is added because it is impossible to predict all of the operating expenses a business will actually require. This extra category is both normal and expected. Finally, total the operating expenses for each reporting period.

Our next step is to calculate the company's future projected profits and cash flows. Subtract the total operating expenses from the total revenues for each reporting period and record this number as the profit or loss for the period. Losses (negative numbers) are reported in parentheses. Make this calculation for the entire 5-year timeline of the proforma. Once the projected profits and losses are tabulated, calculate the company's expected future cash flows (cumulative losses) by summing the monthly profit projections in cumulative fashion. Again, complete these calculations for the entire 5-year proforma period. Now examine the cash flow statement carefully. You will see that the cumulative losses increase at the outset since the company is not generating any profits, peak, then begin to decline toward zero, culminating with positive numbers. The point of the maximum cumulative loss (MCL) is the company's turnaround point (TAP). This is a critical milestone. It represents the projected point at which the company becomes profitable. Make a note of both when the MCL occurs and what the dollar amount is. An equally important determination is the point at which the MCL becomes zero. This is called the break-even point (BEP) and represents the point where all of the cumulative losses are offset by accumulated profits. Again, make note of exactly when the BEP is projected to occur.

You are now ready to make the most important calculation in the entire business planning process: determining exactly how much external capital will be needed to finance your new entrepreneurial business. The following formula is to be used:

$$\text{Total External Financing Need} = (\text{MCL} \times 1.5) + \text{CEC}$$

Simply take the MCL amount, multiply it by 150%, and add the sum total of the CEC. The reason for adding a 50% factor to the MCL is to take into account the business reality that your revenue projections will most likely be higher than the actual revenues and that your operating expense projection will most likely be lower than the actual operating expenses due to the unforeseen operating expenses that all new businesses incur. As before, this is both normal and expected. This resulting amount is exactly how much money you would need in order to properly finance the new business to the point of profitability.

Now that you are armed with the estimated dollar amount of external financing your new business will require, we will discuss some of the different methods available for financing an entrepreneurial business venture. There are essentially three ways to finance a new business venture: equity financing, business loans, or grants.

### Equity Financing, Business Loans, and Grants

Equity financing means exchanging some of the stock in your business for direct investment capital. The outside investment capital can be obtained by entrepreneurs from their own assets, from family and friends, from angel investors (wealthy individuals), or from the venture capital industry. Each of these sources of equity financing has both advantages and disadvantages associated with it. For example, financing a new business with your own funds and those of friends and family will result in the smallest amount of stock being exchanged, but it is a limited source of capital that is likely to run out should multiple rounds of financing be needed. Alternatively, outside financing through angel investors or the venture capital industry represents a much larger pool of investment capital that, for a worthwhile venture, is unlikely to run out. However, the new venture's valuation will be substantially less, meaning that relatively more stock will have to be traded for the same amount of investment capital.

Financing an entrepreneurial business through business loans offers the obvious advantage over equity financing in that there is no exchange of any of the company's stock, and the entrepreneur will maintain 100% control of the operation. However, when a loan is made, debt is created and there is the expectation of repayment of the principal amount of the loan along with interest. Saddling a new entrepreneurial business with debt service, while trying to build positive cash flow, is often a bad idea. As general advice, if your new business venture is projected to achieve its TAP anytime within the first 12 months of operation, financing via a loan is a good idea, because the business will achieve profitability very quickly, and giving away some of the company's stock is a permanent, long-term decision.

The final source of capital for a new business venture is a grant. A grant is a wonderful way to finance a new business venture because there is no exchange of the company's stock and no debt is created in the process. It is basically free money, if you can qualify for it. Most grants are directed toward technology-oriented companies and are very competitive due to their appealing nature. Most granting agencies have Web sites including all of the submission requirements and deadlines associated with the application process. Several examples are:

- Small Business Innovative Research (SBIR): http://www.eng.nsf.gov/ sbirspecs/
- National Institute of Standards & Technology/Advanced Technology Products (NIST/ATP): www.atp.nist.gov

## Organizational Strategy and Management Plan

The final formal section of the business plan includes a description of how the new entrepreneurial business will be legally organized, how it will be grown, who will be hired to manage the business, and how the outside investors will be repaid.

### Legal Organization

There are five choices for legally organizing any business: sole proprietorship, partnership, S-corporation, C-corporation, and limited liability corporation.

Sole proprietorships and partnerships are very similar. Neither of these business entities are corporations and hence offer very little protection for the owners against any liabilities that arise in the normal course of business. The entrepreneur and the business are one in the same for legal and tax purposes. It is very easy and straightforward to organize as either of these two entities. Since they are not corporations, no formal stock has to be created, officers do not have to be appointed, a board of directors is unnecessary, business name searches and approvals are not needed, and in general you do not need to hire an attorney.

S-corporations are great choices for entrepreneurial businesses. First, since they are corporations, there is implied protection from normal liabilities that arise out of conducting business. Second, there is no "double taxation" issue on dividends the owners may receive from time to time. Double taxation means both the company and the individual pay taxes on profits that are generated by the company. But, because they are corporations, formal organization is required. This means establishing a board of directors, appointing officers, creating stock, obtaining a formal tax identification number, and getting approval of your chosen corporate name.

The C-corporation is the most complex corporate entity and has the most flexibility built into it. C-corporations can be both public or privately held corporations. This feature does not exist with any of the other business entities; that is, only a C-corporation can have its stock publicly traded. This flexibility comes at a price. C-corporations' owners are double taxed on any dividends or bonuses the owners may receive. This occurs because the corporation itself pays taxes to the federal government on profits it generates. When the resulting after-tax net profits are distributed to the owners in the form of a dividend or bonus, they are again subject to taxes and must be reported on the owners' personal income tax returns.

There is a relatively new corporate entity called a limited liability corporation (LLC). This is another great choice for an entrepreneurial business. It combines the best features of a partnership and an S-corporation. It allows the flexibility of a partnership when distributing profits to the stockholders and at the same time affords all of the inherent protections of a corporation. LLCs have become very popular with consultancy-type businesses, such as Web design companies.

## Growth Strategies

At some point during the business planning process, you will need to decide exactly what product you want to enter the market with, focus on only that goal, and plan for product improvements and new product introductions at later stages. You should describe your company's future growth and new product plans in this section of the business plan. Try to avoid "planning paralysis," a situation where an entrepreneur is perpetually planning the business and it never gets launched or is significantly delayed to the point where a previously identified market opportunity closes or is filled by a competitor.

## Exit Strategies

If your entrepreneurial business will require outside equity financing of any type, a well-articulated exit strategy is an essential part of your business plan. If it is not clear to investors how they will be repaid for their investment, they may conclude it will never happen and pass on the investment opportunity. Exit strategies can come in the form of an IPO, a sale of the company's assets or stock, or liquidation of the company's assets.

## Management Team

The strength of your proposed management team should not be overlooked. The personality and perceived abilities of the entrepreneurs founding the new business venture is really the only visible asset a business has at its inception. This will be the pivotal issue that will influence an investor's decision to make an equity investment in your new business. Most entrepreneurs have certain areas of relative strength. Maybe you are especially adept at market analysis, preparing financial projections, sales, or human resource management and administration. Whatever your skill sets are, underscore them and seek to complement any areas of possible weakness with others who are stronger in those areas.

## Appendixes

The final section of your new business plan is the appendixes. The first set of documents to include in this section are the resumes of all of the proposed principals of the new business venture. Be sure the resumes are professionally prepared and clearly communicate the relative and complementary strengths of the business team members. In this section also include any of the following items you may have developed, such as:

- Industry reports you have used or cited in the main business plan
- Preliminary marketing or sales material you are planning to use
- Draft office leases
- Intellectual property documentation, including patents, copyrights, and licensing agreements
- Any governmental approvals needed and obtained, such as a business license, tax identification number, business organization documents or third-party billing provider numbers, or codes required
- Any other relevant documentation that you feel is important

## 9.4   SUMMARY

While starting your own business is risky, it can be very rewarding. If you have a novel idea or interesting design project, you should think about entrepreneurship. The process of developing a start-up analysis and a business plan can help you determine whether you are an entrepreneur and whether your product is ideal for building a new business. In this chapter we provided an overview of the business plan process to help you get started. If you think you might want to start your own business you should consider taking classes in business and management and also seeking help from experienced entrepreneurs or organizations, such as incubator centers, which are designed to help fledgling businesses get started.

## REFERENCES

Administrative Office of the U.S. Courts. (February 23, 2001). Bankruptcy Filings Down in Calendar Year 2000. http://www.uscourts.gov/Press_Releases/press_02232001.pdf. (Accessed September 10, 2001.)

Commscan Equidesk. (n.d.). Total Initial Public Offerings on Nasdaq, NYSE, and AMEX. http://www.marketdata.nasdaq.com/asp/sec3ipo.asp. (Accessed September 10, 2001.)

Small Business Administration Office of Advocacy. (August 30, 2001). Economic Statistics and Research. http://www.sba.gov/advo/stats/#general. (Accessed September 10, 2001.)

## FURTHER READING

Amor, D. (2000). *The E-Business (R)evolution: Living and Working in an Interconnected World.* Englewood Cliffs, NJ: Prentice Hall.

   Discusses e-business from planning to marketing and presents the unique aspects of doing business online.

Stevenson, H.H., Roberts, M., and Grousbeck, H.I. (1999). *New Business Ventures and the Entrepreneur,* 5th ed. New York: McGraw-Hill.

   This book uses business cases from Harvard University to explain the entrepreneurial process.

# APPENDIX

Draft Final Report *The Talking Book**
Final Report *Coreware IPv4 to IPv6 Bridge**
Final Presentation *Coreware IPv4 to IPv6 Bridge*

---

*Documents have been reformatted and compressed for this appendix.

# THE TALKING BOOK
## FINAL REPORT

*Submitted to Dr. Maja Bystrom and the*
*Senior Design Project Faculty of Drexel University*

Submitted in partial fulfillment of the requirements for Senior Design Project,
ENGR 493

| | |
|---|---|
| Drexel University<br>*The Talking Book* Senior Design Team<br>David Brouda – Electrical Engineering<br>Kevin Lenhart – Electrical Engineering<br>Team/Project Number 007<br>Talkingbook @ Drexel.org | Revision: 2.0<br><br>Date: May 18, 2000<br><br>Page: 1 of 34 |

1

# 1 ABSTRACT

It is widely known that a parent's reading to a child can improve the child's reading and writing skills and encourage the child's intellectual development. Despite this well-recognized need for adults taking active roles in encouraging children to read, very few products are available that creatively utilize state-of-the-art electronic components to provide efficient, cost effective, and entertaining reading aids. *The Talking Book* will harness the power of digital circuitry and a Radio Frequency Identification (RFID) sensor system to record a parent's voice as he or she reads the book. The child will then hear the story being read back in the parent's voice as he or she pages through the book. *The Talking Book* will allow the concerned parent to encourage a child to read even when he or she cannot be physically present. *The Talking Book* will revolutionize the reading experience for every child through advanced technology with a humanistic touch.

## Table of Contents

# 2 INTRODUCTION

## 2.1 Problem Background

Increases in technology shape society. New technology allows people to quickly and effectively solve problems and complete tasks that may have been impossible only a few years ago. Advances in technology in fields such as communications, healthcare, defense, and entertainment have become such a vital part of our existence that we find it difficult to imagine life without such conveniences. These advances, as well as those yet to come, cannot be possible without the creation of increasingly more effective educational systems. Literacy is one of the most fundamental steps to accomplishing this goal.

Early efforts by parents and teachers to expose children to books and reading have proven beneficial to the children's educational achievement later in life. Introducing children to reading through "shared book experiences," in which classic children's books are read to children by teachers, parents, older children, or audiotapes, have proven effective. As a result of shared book experiences, a study of kindergarten students showed a 10 % increase in the Cognitive Skills Assessment Battery and, therefore, a 10% increase in the number of children being placed in top reading groups in first grade [1]. Other studies have confirmed that shared book experiences provide significant improvement in the development of oral language skills (as measured by standardized tests) [2] and in the early development of language skills. Children who engage in shared reading experiences also benefit from an increased interest in books [3] and greater enjoyment of reading [4]. Given the current state of technology, there is an inadequate amount of high-tech solutions available to aid teachers and parents in shared reading.

In the past, parents and teachers have given children storybooks along with audiocassette tapes or records, which assist the child in reading the book and take the place of the parent's reading. Patents have been filed and products have been sold using cassette tape devices inside books and attached to books. While a cassette can be controlled and indexed, it is hardly an easy-to-use, long-lasting solution to the problem. Other patents have been filed incorporating digital audio devices, some even with devices to determine when a page has been turned in order to play the audio relevant to that page. However, the existing patents call for using "switches" and "inserts" (US5374195) as well as a "capacitive sensor" (US5569868); even "depressible user response buttons" (US4997374). We find that the existing patents are not suitable for application where a child is involved since operation of the device must be simple. Significant technological improvements have been made in the fields of audio digital signal processing and electronic component miniaturization such that the bulky external recording and playback devices are no longer necessary.

We propose to replace these outdated technologies by integrating into the storybook the shared reading experience that cassettes or records once facilitated. This has been accomplished by constructing an advanced shared reading aid from standard "off-the-shelf" electronic components and an advanced sensor system. *The Talking Book* records a parent's or teacher's voice, the high-quality recording then reads back the correct page of the story in the parent's or teacher's voice as the child turns the pages of the book. *The Talking Book* offers an indispensable opportunity to enhance a child's intellectual development and creativity, benefiting from a low cost and high level of convenience to the parent or teacher.

## 2.2 Problem Statement

It has been shown that shared reading experiences between a parent and a child can significantly improve the child's language skills development. Outdated technologies of audiocassette tapes or records have been incorporated into a small number of bulky, expensive, and impractical systems that attempt to assist parents in developing their children's language skills. We propose a cost effective system to encourage and improve language skills development through the proven method of "a shared reading" of classic children's storybooks. *The Talking Book* efficiently integrates state-of-the-art electronic components directly into a classic children's storybook. This system is able to record the parent reading the storybook and sense the appropriate page to play back as the child turns the pages of the book. This allows a parent to successfully promote and contribute to a child's shared reading experience even when not physically present.

## 2.3 Constraints on the Solution

There were several constraints on *The Talking Book*. These constraints can be categorized as: speech storage, RFID efficiency and operating parameters, system operating power, and the total number of pages in a book. In this section we discuss these constraints with respect to the solution we have developed.

The total amount of speech, measured in terms of time, that *The Talking Book* is able to store, is a function of the audio storage device. As will be discussed later, the audio storage device we have selected is the ISD 4004. This device is capable of storage up to 8 minutes of audio, when sampled at 8 kHz. While working with books for the age six-and-under target market (low density of words per page and small number of total pages), eight minutes of audio is sufficient. However, 8 minutes of audio is unacceptable in older target markets, where books will be significantly longer and will have a higher density of words per page. This problem is easily solved by the addition of a second, or even third, ISD 4004 device; these devices can be connected serially so that additional audio can be stored.

The total number of pages is not only limited by the audio storage capacity of *The Talking Book*, but also by the RFID sensor system. The RFID system can only detect a fixed maximum number of tags at any given time. Therefore, we are also limited by this maximum value with regards to the maximum number of pages in *The Talking Book*. The maximum number of RFID tags is a function of operating frequency, processing speed of the on-board DSP in the RFID unit, and the transmitting power of the RFID unit. As RFID matures as a technology, we expect that additional units that meet our requirements will become available which we can integrate into *The Talking Book*.

*The Talking Book*'s system operating power is currently impeded by the RFID component. Due to its high current draw and strict voltage requirements, we are confined to large battery packs for our prototype. The RFID unit we have selected is capable of many more functions than those we are currently utilizing. Thus we feel with a scaled down version, we will be able to overcome these power constraints.

## 3 THE SOLUTION

### 3.1 Method of Solution

In order to determine the solution to the problem, we first determined the limitations on, and the initial "broad" design of, the prototype. Next, we selected components, keeping in mind the constraints in the overall design.

Component selection is important, as it determines the operational characteristics of the design as well as the complexity of the integration that will be required. During the component selection process we decided that *The Talking Book* system would need to be relatively simplistic in order to be easily manufacturable. We selected low power, highly integrated audio processing and control hardware. By wisely choosing the electronic components, we have been able to limit the number of electronic components used in the prototype. We have also elected to use a microcontroller device in *The Talking Book* prototype. This microcontroller will easily facilitate future design updates of *The Talking Book* hardware interface, user interface, or device operation.

After selecting the necessary electronic components the hardware and software design based on the components was initiated. Sections of the hardware design have been simulated in software application packages such as Pspice and HO EESof. All of the microcontroller software written for use in *The Talking Book* prototype was also fully simulated in order to save time in optimization and testing. After simulation, the first prototype was constructed. The design team has evaluated and tested its performance based on the design criteria and has optimized the system. In the future, *The Talking Book*'s electronic design will be miniaturized to facilitate mass manufacturing and to increase the aesthetic appeal of *The Talking Book.*

## 3.2   Design Overview

One key benefit of *The Talking Book* is its ease of use. After the on/off toggle switch is set in the on position, the user of *The Talking Book* needs only to turn the pages of the storybook for the book to operate. If the record/play toggle switch is set in the record mode, *The Talking Book* will begin recording the user's voice as he or she reads the pages of the book. *The Talking Book* uses an Information Storage Device (ISD) chip to store each page of the user's recorded voice. When the user is finished reading a page in the book and turns to the next page, *The Talking Book* recognizes this fact. Radio Frequency Identification (RFID) is used by *The Talking Book* to determine which page of the book is open. *The Talking Book* stops recording and storing the previous page, recognizes the new page, and immediately begins recording the user's voice as he or she reads the new page.

If the record/play toggle switch is set in the play mode, *The Talking Book* will begin playing the story back in a previously recorded familiar voice. *The Talking Book* recognizes to which page the user has opened the book and plays back the appropriate lines of the story. When the user is done listening to a page in the book and turns to a new page, *The Talking Book* immediately begins to play back the new page.

The hardware interaction of *The Talking Book* is illustrated in the System Block Diagram shown in Figure 1 and is described in the following section.

### 3.2.1   Hardware Design

Due to the needs of our Radio Frequency Identification (RFID) sensor system design (see Appendix 1), we have selected an RFID unit from Korteks, as indicated by the top block in Figure 1. The Korteks unit has a simple operation; namely we send the unit a command via an RS-232 interface to poll the RF field for any tags that may be located insides its range. The Korteks unit responds back over the RSS-232 serial port with the value that is associated with the lowest numbered tag located in the RF field. Since we communicate with the RFID unit via RS-232, a microcontroller that supports RS-232 via

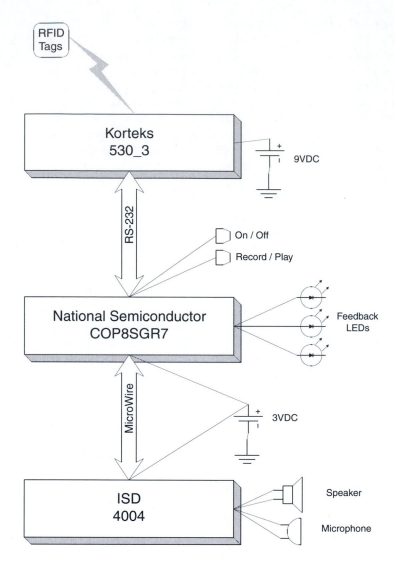

FIGURE 1
*The Talking Book* system
block diagram.

either an on-board Universal Asynchronous Receiver Transmitter (UART) or external UARTs was required.

Since we can directly control a UART when integrated into the microcontroller, we have selected a National Semiconductor COP8SG7 microcontroller. We have programmed the COP8SG7 using an evaluation-programming unit (EPU). This programming unit comes with various software tools for programming and debugging. After receiving and processing the tag value from the RFID unit, we need to determine if the user has selected play or record mode. This is done by looking at the value of the user-selectable switch, which is directly connected to a pin on the microcontroller.

We have developed a simple audio subsystem by selecting an IC that integrates all of our desired functionality into a single chip. This is the lowest block in Figure 1. The ISD 4004 is capable of storing and playing 8 minutes of audio with sampling at 8 kHz through an on-chip ADC, DAC, and memory unit, and directly connect to a microphone and speaker. This permits us to allocate approximately 35 seconds of storage space per page in our 15-page prototype. Using an ISD device will facilitate future performance and operational upgrades to *The Talking Book*. Cascading additional ISD chips will allow us to easily increase the data storage space per page or the number of pages. The ISD 4004 is controlled through a microwire interface SPI (Serial

Peripheral Interface) port. Therefore, we have selected a microcontroller with

**7**

*The Talking Book*
Final Report

an SPI compatible port in order to control the ISD 4004's operations.

### 3.2.2 Software Design

In order to determine user actions with *The Talking Book*, we must program our microcontroller to process data from our RFID device and to control input/output to the book's electronic components. Programming of the micro-controller was done through various software tools provided in an evaluation and programming kit from the manufacturer.

Use input occurs in several forms: power on and power off, turning pages of the book, and selection between record and play modes. Data from the RFID unit is processed after we send a command to the unit. In order to determine when a page of the book is turned, we must continually poll the RIFD unit for the values of the RFID tags currently located in the RF field. The RFID unit returns the tag value associated with the lowest numbered page located in the RF field. We then check to see if the tag value we received is equal to the last one we received or if it is a new value. If it is equal to the last value we received, the page has not changed since the last time we polled; therefore we do not wish to interrupt or start any audio operations. If the tag has changed, we will need to determine if the user has selected to record or play audio through the user-selectable switch, as then we must record or play the audio in our storage device. The programming in the microcontroller must also initialize various portions of the microcontroller, such as the RS-232 port and SPI compatible port, and device timer. The Software Level 0 Data Flow Diagram illustrates logically how the electronic components are controlled by the microcontroller. Additionally, the complete microcontroller assembly code may be found in Appendix 4.

### 3.3 Economic Analysis

*The Talking Book* will fill a market niche that is currently not being filled. As a result, there are no products presently available that could be considered a direct replacement for *The Talking Book*; however, we have identified a few possible competitors.

*The Talking Book* will be economically feasible to develop. A relatively large amount of engineering and material resources are necessary to develop the first prototype; however these costs significantly decrease for additional systems. This decrease in projected cost can be attributed to significant savings on the unit cost of large numbers of electronic components and a reduction in the dedicated engineering support required. In addition, no major capital investments in equip-ment or facilities will be required for the completion of the project.

We expect the cost to produce large numbers of talking books will be quite reasonable, given the well established need for such a convenient reading aid. *The Talking Book* is a "value added" reading aid and provides an unparalleled level of personal appeal.

### 3.3.1 Competitive Matrix

Table 1 shows the competitive matrix comparing *The Talking Book* with its primary competition. After reviewing several high-tech improvements to the standard book, we find the majority of currently available products are not suitable for use by young children. In most cases, the increased product performance facilitated by the technological enhancements is made at the cost of reduced personal affiliation with the product. *The Talking Book* distances itself from the competition by utilizing technology to surpass the performance and functionality of a regular book while simultaneously creating an unmatched level of personal appeal.

Now, with the digital age, books can be downloaded off the Internet onto your computer or PDA (personal data assistant). We believe that this technology is currently too high-tech to reach children age six-and-under. The Rocket eBook™ by Nuvomedia (http://www.nuvomedia.com/) allows the customer to quickly download black and white text and graphics from the Internet directly into a "book-like" reader. The high price tag (starting at $200) and difficulty of use make this product unappealing for young children. Products such as the LeapPad™ by Leap Frog (http://www.leappad.com/) are more suitable for children than e-Books; however they are still much less affordable than a regular book and lack personal appeal.

Unlike a regular book, young children who are just beginning to learn how to read can easily use *The Talking Book* without the presence of an adult. *The Talking Book* also provides the child with a high level of personal affiliation by allowing the book to be read aloud in the familiar voice of a parent, grandparent, or relative.

| Factor | Talking Book | Nuvomedia | Regular Book | LeapFrog |
|---|---|---|---|---|
| Price/Affordability | 3 | 1 | 4 | 2 |
| Ease of Use | 4 | 1 | 2 | 3 |
| Technology | 4 | 4 | N/A | 4 |
| Personal Affiliation | 4 | 1 | 2 | 2 |
| Child Suitable | 4 | 1 | 4 | 4 |
| RANK | 3.8/4.0 | 1.6/4.0 | 2.4/4.0 | 3.0/4.0 |

**TABLE 1**
Competitive Matrix

4-best, 1-worst

### 3.3.2 Market Definition

The target market is the book industry. Overall sales in the book industry for 1998 were $23.033 billion dollars. The industry is then broken down into various segments such as trade, religious, professional, as well as K–12 education and higher education. These segments are displayed with their respective market shares in Table 2.

| SEGMENTS | 1997 SALES | 1998 SALES | %CHG FROM 97 | %CMPD GRTH 97–98 | %CMPD GRTH 92–98 |
|---|---|---|---|---|---|
| Trade | $5453.2 | $6148.9 | 6.5% | 7.7% | 4.0% |
| Religious | 1132.7 | 1178.0 | 4.0 | 5.7 | 4.0 |
| Professional | 4156.4 | 4418.7 | 6.3 | 6.5 | 6.0 |
| K–12 Education | 3005.4 | 3315.0 | 10.3 | 6.3 | 8.0 |
| Higher Education | 2669.7 | 2888.6 | 8.2 | 5.8 | 5.0 |
| Other Categories | 5224.5 | 5084.1 | –2.7 | UTD | UTD |
| TOTAL | $21641.9 | $23033.3 | 6.4% | 6.0% | 5.0% |

**TABLE 2**
Preliminary Estimated Book Publishing Industry Net Sales for 1998 [5]

UTD – Unable to Determine

Trade sales from the previous table are then further broken down into adult hardbound, adult paperbound, juvenile hardbound, and juvenile paperbound. These results are displayed in Table 3. Juvenile hardbound is the category we will focus on since our books will need the feature of a hard cover to support our technology.

| TRADE SEGMENT | 1997 SALES | 1998 SALES | % CHG FROM 97 | % CMPD GRTH 97–98 | % CMPD GRTH 92–98 |
|---|---|---|---|---|---|
| Adult Hardbound (ages >18 years) | $2663.6 | $2741.5 | 3.3% | 6.7% | 3.0% |
| Adult Paperbound (ages >18 years) | 1731.7 | 1908.3 | 10.2 | 9.2 | 7.0 |
| Juvenile Hardbound (ages 0–18 years) | 908.5 | 953.9 | 5.0 | 6.5 | 1.0 |
| Juvenile Hardbound (ages 0–18 years) | 470.3 | 535.2 | 13.8 | 11.8 | 8.0 |
| TOTAL | $5453.2* | $6148.9 | 6.5% | 7.7% | 4.0% |

**TABLE 3**
Additional Preliminary Estimated Book Publishing Industry Net Sales for 1998 [5]

\* This column does not foot. Recopied exactly from the above website. Difference equals $320.9.

The target industry's sales for 1998 represented $953.9 million dollars. This amount will be further broken down into the 3-to-6-year-old age group. Per the Book Industry Study Group, Inc., located in New York City, the number of books purchased for the six-and-under age group is not tabulated. The juvenile segment represents 0-to-18-year-olds. To estimate on the conservative side, we decided to take ½ of target industry sales of $953.9 million to represent age six-and-under juvenile book sales, which represents approximately $477 million. Our goal is to capture 15% of this market in the 5th year of operation, representing $72 million in potential *Talking Book* sales. Since we choose to license *The Talking Book* technology, we estimate collecting a 10% fee for the use of our technology within each book. Of the $72 million, this represents total sales in the 5th year of operation of $7.2 million.

The trend for the book sales industry is increasing. Per Veronis, Suhler & Associates Inc., they expect online book selling to increase. "Online sales totaled $687 million in 1998, which was more than four times the $163 million in 1997." "As online households continue to grow, this will increase spending on books." "In 1998, the average household spent $24 purchasing books online, which is $17 more than 1997." Veronis, Suhler & Associates is predicting sales in year 2003 to be $50 per household and aggregate online spending to be $2.7 billion [6].

The previous tables also show that book industry sales are increasing. Sales rose 5.6% in 1998 compared to 1997. Veronis, Suhler & Associates Inc. also does not foresee E-books having an effect on the "traditional print medium." They state the fact that "online newspapers and magazines have not hurt print circulation despite increased Internet usage." They also comment on the fact that "the book purchasing population tends to be older than the online population with 43% of 1998 book purchases coming from 45-year-olds and older" [6]. This trend may indicate the potential for significant *Talking Book* sales to grandparents and older relatives.

## 3.4   Budget

### 3.4.1   Actual Budget

Our main budget concern thus far has been the cost of the RFID transceiver and tags chosen to complete the sensor subsystem of *The Talking Book*. The RFID technology available at this point in time from commercial off-the-shelf (COTS) vendors is in no way optimized for use in an application like *The Talking Book*. The RFID technology required for use in *The Talking Book* is a greatly reduced (in size, power consumption, read-range, cost, etc.) version of the existing technology. This RFID optimization task is sufficiently involved as to warrant our choice from the available COTS solutions for the scope of this prototype. The increased cost of the RFID transceiver has not been devastating to the progression of our work. Our chosen low cost, highly integrated ISD and microcontroller solution has allowed us to greatly reduce the total number of electronic components in *The Talking Book*. These components have also facilitated the further simplification of the remainder of the prototype design. Table 4 outlines the major expenses incurred to date.

| | |
|---|---|
| Microcontroller Evaluation Programming Unit | $72.00 |
| 3 Reprogrammable Microcontrollers | $54.00 |
| Misc. Electronic Components | $150.00 |
| RFID Transceiver and Tags | $1,537.00 |
| Telephone | $150.00 |
| Office Supplies, Printouts, Shipping, Misc. | $150.00 |
| Total Expenses to Date | $2,113.00 |

**TABLE 4**
Major Expenses Incurred to Date

### 3.4.2   Grants and Awards

We have received the Drexel Family Scholarship in the amount of $1,000. We have also received a silver award in the Entrepreneurship Philadelphia Business Plan Competition in the amount of $250 and are recipients of a Botstiber Engineering Entrepreneurial Award. These funds have been expended on electronic components and material costs incurred to date. To solve our limited funds availability, we will continue to compete for grant funds over the next three months. We are currently preparing submissions for the Drexel University E-Day Business Plan Competition ($1,000 award) and the Collegiate Inventors Competition ($30,000 award). All finances acquired in this fashion will be used to complete the initial prototype of *The Talking Book* and begin the process of miniaturizing *The Talking Book*'s electronics. A summary of the grants and awards received to date is shown in Table 5.

| | |
|---|---|
| Drexel Family Scholarship | $1,000.00 |
| Entrepreneurship Philadelphia Business Plan Comp. | $250.00 |
| Botstiber Engineering Entrepreneurial Comp. | TBD |
| Total Grants and Awards Received to Date | $1,250.00 |

**TABLE 5**
Grants and Awards Received to Date

### 3.4.3   Professional Budget

Appendix 5 shows the projected expenses for *The Talking Book* and Talking Book Inc. through year five of operation. *The Talking Book* is currently in the development stage. The engineers of *The Talking Book* are in the process of patenting the general idea of the book and patenting the sensor system associated with the book's functionality. The financial plan has been formulated to include the initial research and development expenses needed to streamline the technology that will go into each book with future plans of outsourcing the manufacturing and sales processes. The technology is currently available and will be miniaturized to work in *The Talking Book*. We anticipate developing the miniaturization over a nine-month period. This will be our initial ramp-up time starting in June and continuing until February. Until that time, our major need for funds will be in the research and development phase. The June start date, with product development ramping up through February, corresponds with introducing *The Talking Book* in the United States at the annual Toy Fair. The Toy Fair is held in February because the seasonal nature of the toy industry causes many toy retailers to decide what products to carry for the Christmas season during this time period. Unveiling *The Talking Book* at the Toy Fair will provide an exceptional opportunity to generate popular interest within the toy industry.

During the initial research and development phase, personnel, outsourced engineers, and office space are the main expenses that will need funding. The budget will cover salaries for D. Brouda and K. Lenhart. Additional employees will be two outsourced engineers to assist with the project development. Brouda and Lenhart will concentrate on patenting, marketing, and selling the product. The outsourced engineers will concentrate on the miniaturization process and drawings. Employee benefits were estimated at 15% of salaries.

Office space is estimated at $23 a square foot, which is the average rate in the University Science Center area (Philadelphia, PA). We anticipate leasing 5,000 square feet for the corporation. We initially want to stay in the Science Center area since it will lend itself to continued support from the University community. Insurance needed for the corporation is basic business insurance and worker's compensation insurance as required by law for all employees.

Legal fees should be approximately $4,000. Of this, $3,000 represents the patent filing, billed in two increments of $1,500 each, and $1,000 represents the initial incorporation process on the outset. Accounting fees of $1,200 represent the accounting/bookkeeping work and tax filing for the year. All accounting records will be maintained at the corporation's address. All records will be maintained by management and submitted to an accountant at year-end for write-up and tax filings. Records will be maintained in an orderly fashion as to facilitate a smooth transition to audits for three years if needed for additional funding purposes in the future.

Computer hardware will consist of two computers and peripheral equipment. We also anticipate renting computer equipment for the outsourced employees during the development phase. Product development costs are recognized and have been adjusted to reflect the anticipated increase in salaried engineers. Miscellaneous expenses are estimated at 10% of total expense.

Once the research and development phase has been finalized, the major need for funds will swing into marketing the product, Web site, development, professional fees, and salaries. Since the manufacturing process will be outsourced, we will not have manufacturing costs. Various other expenses will include computers and peripherals, telephone and utilities, and miscellaneous expenses. Expenses have been adjusted yearly to reflect estimated increases and the goal of recruiting additional engineers to develop new products and secure additional intellectual property rights.

# 4 DISCUSSION AND CONCLUSIONS

*The Talking Book* is a unique children's book that actually reads to the child. Through improved sensor technology we have designed a children's book which will read to a child with minimal interaction from the child, that is, no other interaction is required beyond a power switch and turning pages. This new technology records the voice of an adult or any other person reading the book and then reads back to the child, recognizing each page as the child reads the book and turns the pages. We believe *The Talking Book* is a personal, as well as an educational, children's product.

*The Talking Book* technology will be marketed to the juvenile book industry, which had total sales of $953 million in 1998. We will target the six-and-under age category, which is estimated to represent at least 50% of the $953 million juvenile market. Our goal is to have *The Talking Book* found on the laps of children across the world.

We anticipate incurring losses for the first nine months of the project lifecycle since this comprises the patent and development phase. After this initial phase is completed, we expect to break even by the second quarter of the second year. Our turnaround point occurs the first month of collecting fees due to the low operating costs associated with our organization, since we are primarily a knowledge-based organization. We will use *The Talking Book* technology as a springboard to developing and marketing new, technologically advanced products. We also plan to continually acquire new staff to develop new products and opportunities.

Upon approaching the problem, we first believed that we could create *The Talking Book* with a hardware-only solution. We quickly discovered that because of the operating requirements of the devices we had integrated, *The Talking Book* became a highly software-dependent project. Due to a lack of exposure to microcontrollers, or any computer engineering coursework or work experience, we battled a learning curve in order to accomplish our goals.

The technical aspects of *The Talking Book* cover a wide range of electrical engineering disciplines, including RF engineering, computer engineering, analog circuit design, digital circuit design, and power engineering. We are both happy and proud to have created a working prototype and look forward to bringing *The Talking Book* through the development phase and into the production, patenting, and product licensing phases.

# 5 RECOMMENDATIONS FOR FUTURE WORK

Due to storage constraints in the audio system, additional ISD 4004 devices in series with the existing device, or new and different devices will need to be added to the hardware design to increase audio storage time.

The RFID portion of *The Talking Book* must be made cost effective. An effort must be undertaken to develop an RFID system for specific use with *The Talking Book*; this development may either be internal or contracted out. The operating frequency of the RFID system should be reevaluated; systems are currently available which operate in the 2.4 GHz range, and research is ongoing with higher operating frequencies. A higher operating frequency would permit the use of a much smaller antenna, and would also yield better performance when multiple tags are present in the RF field. Thus, we should be able to increase the maximum number of readable tags in the RF field. Great consideration should be taken with regard to the RFID's operating power requirements. The current solution's power requirements are not optimal. We only need to excite tags that are relatively close to the antenna so a lower operating voltage and power should be attainable.

# 6  REFERENCES

[1] Brown, Mac H., Cromer, Pamela S., and Weinberg, Sylvia H. "Shared book experiences in kindergarten: Helping children come to literacy." *Early Childhood Research Quarterly*, Volume 1(4), 1986 December, pp 397–405.

[2] Lonigan, Christopher J., and Whitehurst, Grover J. "Relative efficacy of parent and teacher involvement in shared-reading intervention for preschool children from low-income backgrounds." *Early Childhood Research Quarterly*, Volume 13(4), 1998, pp 263–290.

[3] Lyytinen, Paula, Laakso, Marja Leena, and Poikkeus, Anna Maija. "Parental contribution to child's early language and interest in books." *European Journal of Psychology of Education*, Volume 13(3), 1998 September, pp 297–308.

[4] Baker, Linda, Scher, Deborah, and Mackler, Kirsten. "Home and family influences on motivations for reading." *Educational Psychologist*, Volume 32(2), 1997 Spring, pp 69–82.

[5] Association of American Publishers website – http://publishers.org/home.stats/prelim.htm

[6] Veronis, Suhler & Associates – http://www.veronissuhler.com/conbooks/segment.html

# APPENDIX 1: RFID OVERVIEW

## Introduction – RFID (Radio Frequency Identification)

In order to determine the location of pages of a book with respect to the covers of a book, a reliable and efficient sensor technology is needed. Technologies using electrical or mechanical contact methods have been ruled out due to inefficiencies in operations and/or use. Radio Frequency Identification, or RFID, has been selected for use due to its non-mechanical nature and its flexibility.

## Background

RFID has recently come about as a replacement and/or alternative solution for technologies such as barcoding, optical character recognition, voice identification, fingerprint identification, and smart cards.

RFID is quickly becoming the preferred technology in proximity sensors, access control, inventory, and various other sensor applications. While still in its infancy, RFID offers a great deal of flexibility through its ability to operate over a vast range of distances and environments. As RFID matures, a larger number of applications will develop in which the technology will be adopted due to its advantages over more traditional methods.

## Components

An RFID system consists of two basic components: the transponder and the reader. The reader (or read/writer) locates and processes data from the transponder, which is located on the object being identified. The reader transmits a signal through its antenna, which causes the transponder, also referred to as a tag, to respond with unique data. When a tag is within the active field range of the reader, the antenna of the reader receives the signal from the tag, sends it to the reader where it is processed by a programmable device, which can be used to filter and interpret data as needed.

## Operation

The reader transmits a signal at the resonant frequency of the tags. When a tag is within a certain range of the reader's transmitted signal, the tag becomes excited by the reader's transmission and it transmits a data packet. The reader then receives the data packet and decodes the data sent by the tag.

## Properties

RFID systems can operate as full-duplex/half-duplex or sequential systems. In full- and half-duplex systems, the transponder responds to the reader's signal when the reader is transmitting a signal. When a system is operating in this mode, the tag is transmitting data at the same frequency as the reader. Since the reader's signal is much stronger then the tag's signal, the reader would be unable to receive any data from the tag. Therefore, the data transfer between the tag and reader can tag place on either a subcarrier or (sub)harmonics of the reader's transmission frequency.* Sequential systems switch off the reader's transmission briefly, at regular intervals, so that the transponders have an opportunity to respond on the transmission frequency without interference from the reader.

Some operating frequencies commonly found in RFID systems are: 135 kHz, 13.56 MHz, 27.128 MHz, 915 MHz, 2.45 GHz, 5.8 GHz, and 24.125 GHz. The tags used in RFID systems are proportional to the wavelength of their operating frequencies. (For example, tags operating at 135 kHz will be physically larger compared to tags designed to operate at 2.45 GHz.)

RFID systems can employ different options, such as reading and writing to and from tags, authentication and/or encryption of the data communications between the reader and ages, anticollision (the ability to receive data from multiple transponders within the range of the reader), and data processing and manipulation of the transponder data.

## Need

For the RIFD application in *The Talking Book*, we will need to be able to distinguish multiple tags in the range of the reader; for this task our reader must be capable of anticollision. Since pages of a book are extremely close together and will be close to a cover of a book, we need to specify a maximum read range of only several inches. The device must be DC powered at a relatively low voltage (low enough to be battery powered), and cannot consume large amounts of power. For the current design a read-only device is all that is required.

(Additional appendices removed.)

---

* *RFID Handbook: Radio-Frequency Identification Fundamentals and Applications,* Klaus Finkenseller, John Wiley & Son, Ltd.

# *Coreware* IPv4 to IPv6 Bridge
## FINAL REPORT

*Submitted to Dr. Harish Sethu and the*
*Senior Design Project Committee of the*
*Electrical and Computer Engineering Department*
*Drexel University*

*Team Number: ECE-026*

Team Members:

| | |
|---|---|
| Keith Christman | (Electrical Engineering) |
| Adam O'Donnell | (Electrical Engineering) |
| Chayil Timmerman | (Electrical Engineering) |
| Suma Varghese | (Electrical Engineering) |

Submitted in fulfillment of the requirements for the Senior Design Project
May 9, 2001

## ABSTRACT

The explosive growth of the Internet is rapidly exhausting the current address space allocated by Internet Protocol version 4 (IPv4). This problem, coupled with recent "denial of service" attacks, has demonstrated a critical need for a new Internet Protocol. The Internet Engineering Task Force (IETF), in order to rectify the problems associated with IPv4, has recently released Internet Protocol version 6 (IPv6). Due to the radical changes deemed necessary by the designers, the two protocols are not interoperable. The size of the Internet and the existence of legacy software inhibit a speedy migration from IPv4 to IPv6. In fact, this transition will take years, if not decades, to complete. The continued growth of the Internet hinges upon the adoption of IPv6. So as not to segregate the Internet into two separate islands, one based on IPv6 and the other on IPv4, there is a strong desire for a bridge to span the divide.

This report describes how the design team was able to develop a network level device that will allow for communication between the IPv4 and IPv6 networks. This product was developed using recent advances in digital hardware design, allowing for a high speed, low cost product. The solution described in this report will be delivered in the form of an Intellectual Property block, or coreware. This design would be licensed to hardware production customers to embed in their network level devices. The development of this bridge between the two network protocols will ease the transition from IPv4 to IPv6. Ultimately, this coreware solution would facilitate an accelerated migration to an Internet that provides improved security, quality of service, and ease of configuration.

## Table of Contents

## List of Figures

## I. INTRODUCTION

### A. Problem Background

In recent years, the growth of the Internet has been dramatic. The drive to connect additional nodes to the network is constantly increasing, while the amount of addressable space in the current protocol scheme continues to diminish. While it may seem that the 4.3 billion addresses provided in IPv4 (Internet Protocol version 4) is rather extensive, a large number of these addresses have already been delegated to large-scale users or reserved for highly specialized networking purposes. Therefore, the number of new nodes that can be connected to the Internet under the current addressing and net-working scheme is shrinking rapidly [1].

On top of the address space issue, there is the increased amount of load placed upon routers and the network infrastructure required to process IPv4 packets. There is excessive complexity introduced by several unnecessary parameters, specifically error-detection codes. Security concerns, such as source address verification, have recently become major issues. These prob-lems have been addressed by the development of a new Internet Protocol, known as IPv6 (Internet Protocol version 6) [2].

IPv6 provides solutions to the majority of the issues discussed. Addresses have increased from 4.3 billion to $3.4 \times 10^{38}$ possibilities and are assigned through a hierarchal scheme. These addresses are self-derived by each node connected to the network. Security issues that arose due to unverified source addresses are reduced by the requirements of the hierarchal addressing. Redun-dancies, such as error-detection coding in the IP header, are removed, while new parameters are added that allow for extremely rapid routing. These changes have allowed IPv6 to be a much more secure and efficient protocol to power the Internet [2].

## B. Problem Statement

The migration from IPv4 to IPv6 will not be easy. For an operating system to support IPv6, many services need to be upgraded. These services range from the base TCP/IP stack to every application that does any form of networking using the Internet Protocol. Host-level solutions are possible, but in many instances, not applicable. In the case of modern desktop computers where operating systems are regularly updated by the vendor, the migration for the end users would be fairly transparent. This sort of upgrade becomes economically prohibitive for legacy systems whose operating systems have been abandoned for years. This results in the inoperability depicted in Figure 1.

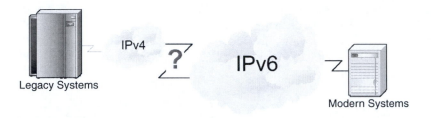

**FIGURE 1**
The problem with
communication between
IPv4 and IPv6.

Our proposed solution is to provide connectivity through a network level device, as seen in Figure 2. The purpose of this device would be to allow the host to converse in IPv4, but would provide some method of communication to nodes on an IPv6 network. The end product will be a fully tested HDL (Hardware Description Language) description of the device, which would be sufficient for major hardware vendors to embed in their systems. HDL descriptions, or coreware, allow system integrators to embed designs into their large-scale devices with ease.

Device development using coreware solutions from multiple vendors is now standard industry practice. Besides offering a cheaper alternative, coreware solutions allow a hardware design to adapt readily to changes in technological constraints. For these reasons, we have chosen to present our solution as a coreware product.

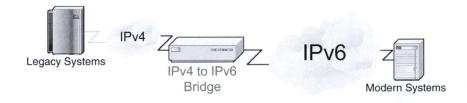

**FIGURE 2**
IPv4 to IPv6 network
bridge.

The device will directly translate the IPv4 headers to IPv6 headers through pipelined logic. The payload of the packets will then follow the new headers, unaltered by the calculations. If these translation actions are performed in a standards-compliant manner, then legacy hosts using the older networking protocol would be capable of communication over the newer IPv6 network.

## C. Constraints on the Solution

It is easier to examine the numerous constraints that have been placed upon our solution by dividing them into two broad categories: *Industry Standards* and *Technical End-Product Constraints*. *Industry Standards* constrain our device to conditions of interoperability, based upon the standard networking protocols. The *Technical End-Product Constraints* are the typical economic considerations usually involved in an engineering project.

Well-known and accepted RFCs, or Requests For Comments, set the majority of the Internet Standards to which we must comply [3, 4]. The design must follow the packet header formats and field values provided in these RFCs. The remaining constraints are imposed by economics: cost, time to develop, and complexity. Issues such as gate consumption, chip-level memory size, data path bandwidth, and end-to-end latency translate into system complexity and time to develop. The architecture selection and methods for translation were governed by these chip level economic constraints.

## II. STATEMENT OF WORK

### A. Method of Solution

We propose to implement the IPv4 to IPv6 Bridge in hardware by utilizing HDL, specifically VHDL (Very High Speed Integrated Circuit Hardware Description Language). Improvements in HDL allow for complex algorithms that were originally constrained to software implementations to be placed completely in the hardware domain. In recent history, high-level HDL design tools have become the accepted methodology for logic designers [7]. The remainder of this section will be devoted to the research and design of the network level device.

Research was first undertaken to determine the actual definitions of what constituted an IPv4 and IPv6 packet. Most of these answers were found in RFCs. These documents form the standards upon which the vast majority of Internet protocols are based. The RFCs of particular interest were RFC-791 and RFC-2460, on IPv4 and IPv6, respectively [3, 4]. Upon gaining a firm understanding of the two packet formats, the design team was able to develop a pipelined architecture to translate between the two protocols.

A translation scheme of this nature was described in RFC-2765, entitled "Stateless IP/ICMP Translation Algorithm (SIIT)" [8]. This RFC was then implemented in the form of a Windows NT device driver, by a team at the University of Washington [9]. Using RFC-2765 and the report from the University of Washington team, pipelined algorithms have been developed to handle the four different possibilities that may occur while translating between IPv4 and IPv6 [8, 9]. We have developed flow charts to detail the translation scheme for each of these header fields found in Appendices K–N. Unfortunately, neither document provides an acceptable solution for translating the address fields contained within the packets.

Research and development was performed in the laboratories of Drexel University's Electrical Engineering Department in Commonwealth Hall. The design team used ModelSim for the compilation and testing of the design. ModelSim is provided to the Drexel University Undergraduate Electrical Engineering community as part of a student's laboratory fees. ModelSim provides an all-encompassing simulation tool. With the use of this tool and an extensive simulation scheme, the design was tested for both functional and logical faults. Test data was collected from IPv4 and IPv6 networks to provide the simulated bridge with real world stimulus.

## B. Alternative Solutions

As mentioned before, host-level solutions already exist on the market today. These solutions consist of software applications that modify the operating system's TCP/IP stack enabling communication via IPv6. Additionally, any application that utilizes IPv4 must be updated for communication over the new protocol. This becomes cost prohibitive on legacy systems where the original vendors are no longer interested in supporting their product. The reality of expensive code upgrades became apparent during the rush to make systems Y2K compliant. Industry spent billions of dollars to ensure that their systems would operate into the next millennium. Software solutions have the major advantage of being inexpensive and having a rapid development time. These software solutions could be easily applied to systems that are continually upgraded, but would neglect the aforementioned legacy systems. Because of this incompatibility, this method of solution was eliminated from consideration.

It is also possible to develop a software-oriented network solution. This would be based upon an off-the-shelf microprocessor with custom embedded code written for the purpose of IPv4 to IPv6 translation. The device would be implanted inside a standard router. This alternative carries many of the same advantages as does a hardware-based network level solution, including low cost of rollout. It is important to mention, however, that there is a great deal of added latency introduced by the overhead associated with a microprocessor. A custom hardware-based solution would provide a more cost-effective method of delivering high data bandwidth and low conversion latency to the translation problem.

## C. Progress toward Solution

### 1. Architecture Design

The architecture is divided into two pipelines: IPv4 to IPv6 translation and IPv6 to IPv4 translation. The pipeline architecture utilizes a 32-bit data bus. The network interfaces are directly connected to this 32-bit data bus. This value was chosen because the source and destination addresses in IPv4, which are critical to the translation process, are 32 bits wide.

The design consists of several functional logic blocks. A decoding block was developed to parse the incoming 32-bit words into the separate fields that comprise the packet headers. The structure of the headers can be seen attached in Appendix J. This decoding block separates the individual fields, such as *VERSION*, *IHL* (Internet Header Length), and *TOS* (Type Of Service), and passes them to the translation logic. It allows the follow-on logic to work with the input words passed to our device as individual fields.

The next level of logic consists of translation blocks that convert the various fields from IPv4 to IPv6 specifications and *vice versa*. The translation algorithms of these logic blocks are depicted in the flow diagrams found in Appendices K–N. The most complex translation block is the address translator. As a placeholder solution, we have decided to utilize a four-element lookup table for the address translation. This temporary solution allows for the continued design of the pipeline architecture without an elaborate address translation scheme. This allowed the group to move through the development of the design quickly and granted us additional time to incorporate other features. Using the current configuration, the router manufacturer would have to specify a list of compatible IPv4-IPv6 address-pairs to be handled by this device.

The encoder is the last of the major logic blocks. The encoder performs the complementary function of the decoder. This block accepts the translated fields and packages them into 32-bit words, which are then sent to the network. The fields are aligned as shown in the header formats found in Appendix J.

Once the modules were designed and tested, it was possible for the design team to focus attention on the construction of the pipelines. Two major design issues became apparent due to the inherent differences between the protocols. The first issue is that IPv4 headers can be of variable size. The second is that IPv4 headers are, on average, smaller than their corresponding IPv6 headers. The former problem was addressed with a variable length buffer space for both the packet data and the individual header fields. Buffers are blocks of temporary storage space that hold data statically for multiple clock cycles. Buffering is required because multiple packets may be in our device at any given time. Without state-saving buffers, there is a risk of data corruption due to the collision of sequential packets. The latter problem was addressed by controlling the rate at which new packets arrive at our device. Delaying incoming packets provided the additional clock cycles the pipeline needs for the encoding of the longer IPv6 packet header. The pipeline structure was designed in a way that minimizes idle time in the pipeline, while performing accurate and reliable packet translation. A more in-depth explanation of the IPv4 to IPv6 pipeline is provided in Appendix B.

The design of the IPv6 to IPv4 pipeline encountered fewer roadblocks due to the knowledge gained from the first undertaking. The IPv6 packet header is inherently longer, thus providing more clock cycles for translation. This eliminates the need to delay packets in buffers before entering this pipeline. The major design considerations encountered were related to the variable number of stacked headers in IPv6. This affects our buffering scheme and translation timing. Appropriate changes were made to the pipeline structure to accommodate these issues. A more in-depth explanation of the IPv6 to IPv4 pipeline is available in Appendix C.

## 2. ICMP Implementation

ICMP (Internet Control Message Protocol) is required to alert hosts utilizing the bridge of possible error conditions that may occur. Support for this protocol was added to our IPv4 to IPv6 pipeline architecture. Since the device is intended to be placed on the edge of a small IPv4 network it is only essential to inform the IPv4 hosts of any packet errors. For this reason, ICMP was only incorporated for IPv4 hosts. Adopting this protocol into our device required considerable changes to our pipeline structure, such as error detection and packet storage. A more in-depth explanation of the ICMP implementation can be found in Appendix D. The adoption of this protocol and its functionality has made our device a far superior product.

## 3. Simulation

The senior design team generated individual test vectors that represent both fragmented and unfragmented IP traffic. This was done as initial testing to discover any inherent logical errors. Simulation output timing diagrams for both pipelines can be seen in Appendix O. These test vectors do not fully verify our design and a more comprehensive simulation scheme was developed and implemented. To aid in this complete testing, a software tool has been developed to intercept IPv4 packets from an operating network. An explanation of this software tool is provided in Appendix P. These packets were translated using a software model isolated from the pipeline VHDL design. Packets from this software model were compared to those processed by the pipelines. A more detailed explanation of the simulation effort is provided in Appendix O.

## III. PROJECT MANAGEMENT TIMELINE

In order to properly gauge the progress of the development effort, a Gantt chart was created in the fall and approved by the group, which is attached in Appendix Q. The chart details the major stages of the engineering effort, including time expenditures incurred for documentation and presentation of the results. All deadlines have been met at or before our projected date. Specifically, the pipeline architectures were completed before our proposed deadline. This allowed for future efforts in ICMP packet generation and provided more time for exhaustive simulation testing. Our Gantt chart was modified to reflect the changes. A final version of this chart has been provided in Appendix S.

## IV. ECONOMIC ANALYSIS

### A. Cost Considerations

The actual costs of what we propose to deliver are relatively low as compared to similar engineering endeavors completed in the private sector. The only actual cost incurred by the project is that of tuition, and the benefits associated therein. Associated with tuition is the full use of Drexel's facilities, including the use of high-powered HDL design and testing tools, along with office, meeting, and laboratory space. Appendix T shows the estimated costs, to the design team, that are covered by their tuition dollars. The estimated total is $128,000. It is harder to quantify the tangible cost incurred in this design project. Therefore we must examine the opportunity cost associated with this project. The opportunity cost for the team totaled $72,000. This assumes that each member, instead of working on the project, would work 20 hours a week as an engineer with an average hourly rate of $25 per hour.

Appendix T presents a budget based on what it would cost a comparable company operating in the private sector. An additional $551,000 would be needed to cover the costs of salaries, office space, and overhead.

### B. Sales and Profit Considerations

As stated before, the end product of the development effort will be an HDL description of an IPv4 to IPv6 translator. This HDL description would be licensed to the hardware vendor to implement in their ASIC-based design. Following standard licensing agreements, the license would be sold for a base rate of approximately $25,000 to $50,000. The license would also stipulate that a portion of profits based upon the number of end product devices produced, using our design, would be allocated to us.

## V. SOCIETAL AND ENVIRONMENTAL IMPACT ANALYSIS

It is difficult to analyze the "benefit" of a product that will increase the availability of the Internet without making the basic assumption that cheap and easily accessible data communications is a benefit to society as a whole.

From this point further the assumption will be made that universalizing access to the Internet would be a gain for society.

Migration of IPv4 to IPv6 would allow for more hosts to be connected to the Internet. Due to the new protocol, routers will become cheaper and display lower routing latencies. As a result, access costs to the backbone would be reduced. These costs can then be passed on to the multitude of consumers, ranging from private sector to governments.

Although the true "paperless" office may never be realized, this technology would further reduce the cost of internetworking, where e-mail has replaced hardcopies of memos. A positive environmental impact would be realized by the reduction of the consumption of paper products and the subsequent waste generated by their disposal.

## VI. CONCLUSIONS AND FUTURE WORK

The design of the IPv4 to IPv6 Bridge was a successful endeavor for the design team. The expectations set forth at the beginning of the design were very realistic considering the work completed. Address translation for the design was an open-ended issue and was not completely developed. Instead, the design team focused on integrating ICMP into the design to incorporate additional functionality. Invaluable experience was gained designing the pipeline architectures and working through various roadblocks that were encountered. Significant architecture changes and implementations had to be incorporated into the design to accommodate for issues that were unforeseen. Some of these issues were not addressed in the design and are a matter of future research. The four-element lookup table allowed for the rapid design of our device without the overhead related to a complex address translation scheme. Future work could be directed toward this area of research. A third pipeline would have to be constructed to capture DNS (Domain Name Server) packets. These packets could be utilized to create a dynamically updated list of compatible address pairs. This configuration would require the implementation of a dictionary chip or searchable memory core. The DNS engine would update this searchable memory core with addresses that are to be processed by the pipeline architectures. The translation logic in the pipeline would then utilize this searchable memory to find the matching entry and the corresponding address translation. The design currently has an address interface that could be utilized with a searchable memory chip. Also, the design has an expandable buffering scheme, which would allow the pipeline to be expanded. This may be necessary due to the fact that more clock cycles would be needed to translate addresses inside the searchable memory chip. Currently, the four-element table translates addresses in only one clock cycle. It is expected that complex searching algorithms inside the searchable memory module would take far more clock cycles to complete the translation. As an added feature, ICMP could be integrated on the IPv6 to IPv4 pipeline architecture. This would add ICMP functionality for IPv6 hosts. This could be done in a similar fashion to the design team's implementation for IPv4 hosts.

## VII. REFERENCES

[1] IPv6: The Next Generation Internet! (10/18/2000). IPv6: Networking for the 21st Century. http://www.ipv6.org. [11/07/2000].

[2] IP Next Generation Overview. (05/14/1995). IP Next Generation (IPng). http://playground.sun.com/pub/ipng/html/INET-IPng-Paper.html. [11/07/2000].

[3] RFC 791 — Internet Protocol. (09/1981). Internet RFC/STD/FYI/BCP Archives. http://www.faqs.org/rfcs/rfc791.html. [09/26/2000].

[4] RFC 2460 — Internet Protocol, Version 6 Specification. (12/1998). Internet RFC/STD/FYI/BCP Archives. http://www.faqs.org/rfcs/rfc2460.html. [09/26/2000].

[5] RFC 792 — Internet Control Message Protocol. (09/1981). Internet RFC/STD/FYI/BCP Archives. http://www.faqs.org/rfcs/rfc792.html. [09/26/2000].

[6] RFC 2463 — Internet Control Message Protocol (ICMPv6) for the Internet Protocol Version 6 (IPv6) Specification. (12/1998). Internet RFC/STD/FYI/BCP Archives. http://www.faqs.org/rfcs/rfc2463.html. [09/26/2000].

[7] J. Bhasker, *A VHDL Primer,* Upper Saddle River: Prentice-Hall, 1999.

[8] RFC 2765 — Stateless IP/ICMP Translation Algorithm (SIIT). (02/2000). Internet RFC/STD/FYI/BCP Archives. http://www.faqs.org/rfcs/rfc2765.html. [10/03/2000].

[9] The Design and Implementation of an IPv6/IPv4 Network Address and Protocol Translator. (06/1998). Brian N. Bershad [Personal Homepage]. http://www.cs.washington.edu/homes/bershad/Papers/USENIX98/nap.html. [10/03/2000].

# APPENDIX A
## ALTERNATIVE DESIGN SOLUTIONS

Two overall architecture design schemes were considered as the primary design paths for the proposed bridge: the first being a Central Processing Unit (CPU) architecture, and the second being a Pipeline architecture. The CPU architecture is similar to that of a personal computer, where all tasks are scheduled in the time domain and executed in sequence. As the name implies, the Pipeline architecture allows the data being processed to flow through the system unobstructed, with many operations being performed in parallel. Each method has its advantages and disadvantages. The CPU-based solution is usually less complex and easier to implement as compared to a Pipeline. It is important to note that Pipeline architecture is considerably faster as measured by the processing time in latency-sensitive applications.

The CPU architecture was deemed a less optimal solution for this application. The CPU method's packet translation time would be directly related to that of the bandwidth and throughput of the memory bus. At a minimum, each packet that was to be translated would have to be written to memory and subsequently read out of memory. This design path would also depend on a vast amount of core memory, which is typically large and expensive. It is also very similar to the current microcomputer-based software solutions. Since low-latency and system throughput was one of the critical design goals, the CPU-based solution was eliminated from consideration.

# APPENDIX B
## IPv4 TO IPv6 PIPELINE CONSTRUCTION
## (TECHNICAL CONSIDERATIONS)

Several characteristics of the IPv4 and IPv6 protocol headers and their differences make the IPv4 to IPv6 conversion a very intriguing design problem. The IPv6 packet header is much longer, consisting of either 40 bytes or 48 bytes of data, depending on the fragmentation state. Conversely, the base IPv4 header is only 20 bytes. IPv4 allows for variable length options fields that may exist in the header. These fields are ignored in the IPv6 domain. Incorporating these differences while maintaining a pipeline structure, which allows data to flow unobstructed through our device, presented an extremely difficult design problem. Additionally, the packets have to be processed without adding excessive delay due to our translation scheme.

Most of the work for our design was oriented toward the difficult pipeline considerations necessary for our device. The pipeline, shown in Appendices F and G, is documented for both fragmented and unfragmented packets. These diagrams show when each line of the IPv4 header is captured, what calculations are done in each clock cycle, and when the IPv6 output is encoded. The first decision that had to be made in the design process was determining the clock cycle during which the device would begin providing output. The decision was dominated by the fact that a missing address in our lookup table would initiate an error, terminating the output of the packet. As can be seen in our pipeline diagrams, the last address is calculated on clock 7. It would have been possible to start outputting the IPv6 packet at clock 8, but our device adds two inherent clock cycles latency due to propagation delay between logical units. This forced the design to begin output on clock cycle 10. A central control logic block was added to initiate the start of the encoding process. This block consists of an internal counter incremented by each clock cycle. Once the counter reaches the desired output time, the encoder begins to generate the output words. The device is scheduled to provide output on clock cycle 10, but one clock cycle is lost due to the propagation delay associated with the begin encoding signal. For this reason we must set this signal by clock cycle 9. This scheme starts the output of the IPv6 packet on the 10th clock cycle.

The next problem that had to be resolved was the fact that a variable amount of delay had to be added to incoming packets. This is necessary because the IPv6 packet header requires much more time to output than to read the corresponding IPv4 header. If IPv4 packets arrived at too high of a rate, the pipeline would overflow and packets would be lost. After much research and experimentation, it was determined this could be done with a "front-of-the-line" ring buffer. This buffer is controlled by a read and write pointer and consists of registers for fast access. The data passed into our device is buffered if the incoming data pin is held high. The read pointer is delayed for a specified amount of time between packets. This allows the device to maintain the pipeline integrity by delaying data before it is passed to the decoder. The number of options found in the previous IPv4 packet determines the number of clocks of delay needed. Luckily, IPv4 has a header field that provides the number of options in a packet. This value accesses a lookup table to determine how many "front-of-the-line" buffer stalls are necessary. The values for the number of "front-of-the-line" buffers, along with the number of data buffers (which will be discussed later) are found in Appendix E. Because of the inherent two cycles of propagation delay these numbers are actually two clock cycles short of the actual delay for the data. Fragmentation causes the device to need two more front buffers because of the additional two lines of IPv6 output that form the fragmentation header. Finally, we determined that under continuous packet traffic our device's ring buffer would eventually fill up, causing some packet failures. It was decided the device would work under a consumer/producer model for packet supply to our device. When our buffer is empty the producer, or router, will provide data. When the buffer fills, we signal the producer to stop providing data until we can catch up to the current traffic.

The final major issue confronted in the pipeline design was also related to buffering, but this time for the actual packet data. The data portion of the packet is passed on a single internal bus in the device. For this reason each data word needs to be placed in a buffer, or holding logic, until it is to be encoded by our device. The number of options in the current packet and the fragmentation state determines the length of the buffer needed in the data path. A lookup table is utilized to provide the correct number of data buffers for the device. The number of necessary data buffers for all conditions is attached in Appendix E. A variable length data buffer was specifically coded for the task. However, it was discovered that this innovation could possibly corrupt data. Since two packets worth of data can be in flight in our device at a time, it is possible to have data corruption in this data buffer path. The design was then modified to include a dual path variable length data buffer scheme, which solved this difficult problem. Two banks of buffer space are found in our device. The control logic continually switches between these two paths to extract and insert data into our pipeline data buffers. This novel solution eliminates the chance of data corruption within the pipeline.

# APPENDIX C
## IPv6 TO IPv4 PIPELINE CONSTRUCTION
## (TECHNICAL CONSIDERATIONS)

The IPv6 to IPv4 pipeline was the easier of the two pipelines to construct. However, there were a few technical challenges that made its implementation a nontrivial task. The IPv6 protocol allows for an undetermined number of extension headers following the main header. These extension headers, with the exception of the fragmentation header, are ignored in accordance with the IETF (Internet Engineering Task Force) RFC (Request for Comments) 2765 *Stateless IP/ICMP Translation Algorithm*. There is also the issue of a CRC (Cyclic Redundancy Check) calculation that is needed for the IPv4 header. The aforementioned issues increased the complexity of the original encoder and decoder blocks.

The decoder block at the beginning of the pipeline was modified so as to remove unused information contained within the extension headers. This action has the effect of holding the state of the pipeline constant until the extension headers are flushed from the input. The existence of an extension header is determined by examining the *Next Header* field in the IPv6 header. The decoder classifies the data following the header into one of three classes: data, unused extension headers, and fragmentation headers. Since the fragmentation header is considered an extension header, it can be present anywhere between the end of the IPv6 header and the beginning of the data. Its placement in the packet has a negative impact on the actual number of clock cycles for translation. The pipeline cannot begin translating the *Fragment Identifier*, *Fragment Offset*, or *Flags* fields in the IPv4 packet until it is known for certain if the incoming IPv6 packet is fragmented. The translation start time is delayed until the start of the data portion because the presence of a fragmentation header dramatically changes the values of the IPv4 *Protocol and Total Length* fields. These requirements dictated a drastic overhaul of the initial design of the decoder block.

The start signal for the encoding process was set when the pipeline received the second data word of the input packet. It was decided that this would be the best place for the signal to be set because at this time all the necessary IPv4 fields would be completely translated and ready for encoding. For a detailed examination of the IPv6 to IPv4 pipeline's timing diagrams please reference Appendices H and I. It takes one clock cycle for the start signal to propagate to the encoder and for the encoder to begin the output of the translated packet.

The *CRC* field present in the IPv4 header utilizes a 16-bit Internet Checksum. This is performed by taking the one's complement of the IPv4 header divided into 16-bit blocks with the *CRC* field initialized to zeroes. In order to reduce the size of the adder tree, we first performed a 32-bit Internet Checksum on the IPv4 packet, with the *CRC* field initialized to zeroes. This 32-bit number was then split into two separate 16-bit numbers, which were then added in a one's complement fashion. It can be proven that this approach is mathematically valid. By embracing this methodology for calculating the *CRC* there was no unnecessary latency injected into the pipeline.

# APPENDIX D
# ICMP TECHNICAL CONSIDERATIONS

There are three error conditions that are supported by our ICMP implementation. They are host unreachable, net unreachable, and hop limit expired. The first two errors occur if the source or destination address is not present in the lookup table for our device. The third occurs when the Time to Live for the packet is exceeded and has to be dropped from the network. The translation blocks have been configured to alert the system when any of these error conditions have occurred. New logical blocks were added to detect and process these errors within the pipeline architecture. ICMP packet generation requires that the IP Header and first 64 bits of the packet payload must be transmitted in the error packet. To support this feature a small memory module was added to the design to buffer offending packet information. When an error is detected the packet that is currently being processed will be saved to memory for future use. The error logic also prevents transmission of the converted offending packet onto the outgoing network interface. Generation of the ICMP packet takes more clock cycles than a standard packet translation. Also, bursts of error generating packets could potentially overflow the allocated memory module in our design. For these two reasons, the pipeline is stalled to prevent new packet translations from occurring during ICMP packet generation. This required a complex signaling strategy along with considerable changes to the front buffering scheme.

A new encoding block was designed to generate the ICMP packet. Each field of the IP header as well as ICMP header was detailed in RFCs 791 and 792. The packet header formats can be seen in Appendix J. The only two fields of interest were the checksum calculations, present in both the IP and ICMP headers. These were calculated using the 16-bit Internet Checksum calculation. The encoding block outputs the IP and ICMP headers in 32-bit words, just as was done for both pipeline architectures. The payload for the ICMP error packet is the offending IPv4 packet header and 64 bits of data. The encoder acquires this data from the allocated memory module previously explained. The encoder then outputs this data in design compliant 32-bit words to the network interface with a valid ICMP packet out signal. To conclude the ICMP packet generation, a signal is sent to the pipeline architecture to resume standard packet translation.

# APPENDIX E
# PIPELINE CHARACTERISTICS (IPv4 TO IPv6)

**Necessary Control Logic:**

Start Encoding Signal: This signal is set at the end of cycle 8. To begin output at cycle 10.

Front End Buffer Control: Allows bypassing of front of the line buffers and is based on the previous packet options and fragment state.

Data Buffer Control: Allows bypassing of data buffers and is based on the current packet options and fragment state.

**Fragmented Case:**

| Options | Start encoding | Front of the line buffers | Data buffers |
|---------|----------------|---------------------------|--------------|
| 0 | 8 | 7 | 13 |
| 1 | 8 | 6 | 12 |
| 2 | 8 | 5 | 11 |
| 3 | 8 | 4 | 10 |
| 4 | 8 | 3 | 9 |
| 5 | 8 | 2 | 8 |
| 6 | 8 | 1 | 7 |
| 7 | 8 | 0 | 6 |
| 8 | 8 | 0 | 5 |
| 9 | 8 | 0 | 4 |
| 10 | 8 | 0 | 3 |

**Unfragmented Case:**

| Options | Start encoding | Front of the line buffers | Data buffers |
|---------|----------------|---------------------------|--------------|
| 0 | 8 | 5 | 11 |
| 1 | 8 | 4 | 10 |
| 2 | 8 | 3 | 9 |
| 3 | 8 | 2 | 8 |
| 4 | 8 | 1 | 7 |
| 5 | 8 | 0 | 6 |
| 6 | 8 | 0 | 5 |
| 7 | 8 | 0 | 4 |
| 8 | 8 | 0 | 3 |
| 9 | 8 | 0 | 2 |
| 10 | 8 | 0 | 1 |

Coreware IPv4
to IPv6 Bridge
Final Report

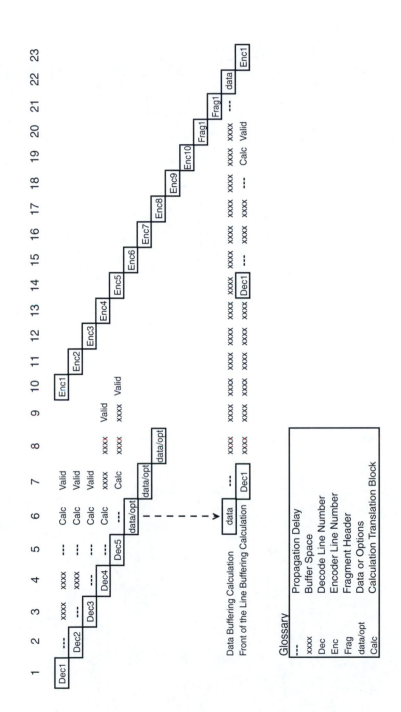

IPv4 to IPv6 Pipeline Timing Diagram for Fragmented Packets

Glossary

| | |
|---|---|
| --- | Propagation Delay |
| xxxx | Buffer Space |
| Dec | Decode Line Number |
| Enc | Encoder Line Number |
| Frag | Fragment Header |
| data/opt | Data or Options |
| Calc | Calculation Translation Block |

Coreware IPv4
to IPv6 Bridge
Final Report

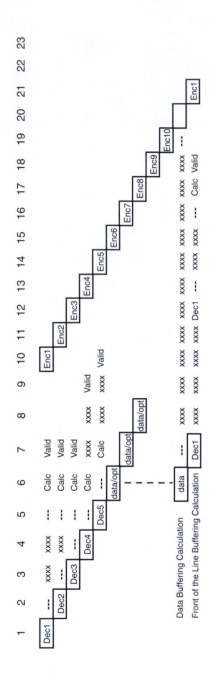

IPv4 to IPv6 Pipeline Timing Diagram for Unfragmented Packets

Glossary

| | |
|---|---|
| --- | Propagation Delay |
| xxxx | Buffer Space |
| Dec | Decode Line Number |
| Enc | Encoder Line Number |
| Frag | Fragment Header |
| data/opt | Data or Options |
| Calc | Calculation Translation Block |

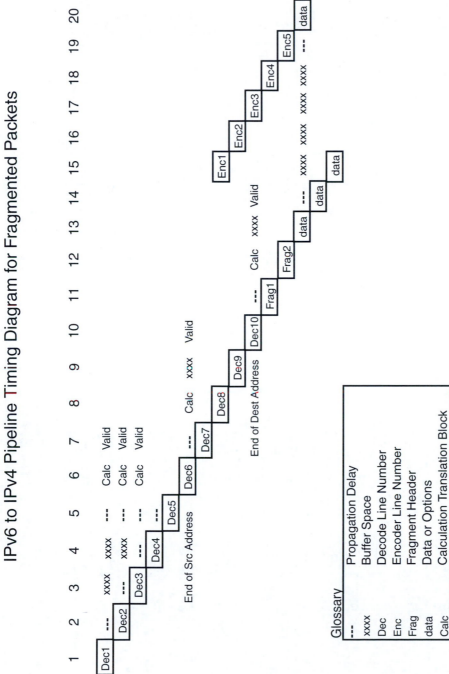

IPv6 to IPv4 Pipeline Timing Diagram for Fragmented Packets

Glossary

| | |
|---|---|
| --- | Propagation Delay |
| xxxx | Buffer Space |
| Dec | Decode Line Number |
| Enc | Encoder Line Number |
| Frag | Fragment Header |
| data | Data or Options |
| Calc | Calculation Translation Block |

## IPv6 to IPv4 Pipeline Timing Diagram for Unfragmented Packets

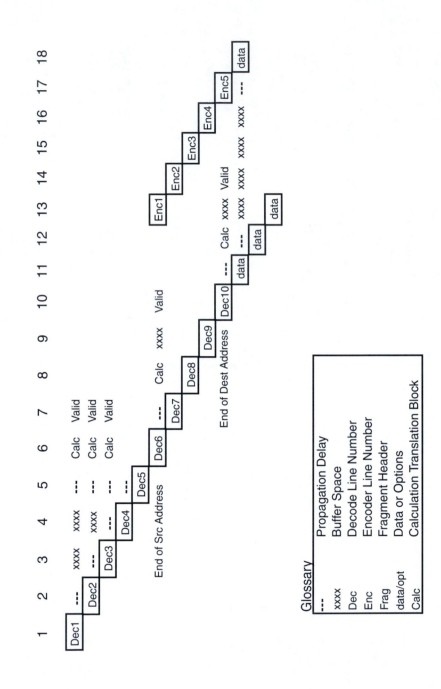

Glossary

| | |
|---|---|
| ---- | Propagation Delay |
| xxxx | Buffer Space |
| Dec | Decode Line Number |
| Enc | Encoder Line Number |
| Frag | Fragment Header |
| data/opt | Data or Options |
| Calc | Calculation Translation Block |

**IPv4 Header Structure**:

```
 0                   1                   2                   3
 0 1 2 3 4 5 6 7 8 9 0 1 2 3 4 5 6 7 8 9 0 1 2 3 4 5 6 7 8 9 0 1
+-+-+-+-+-+-+-+-+-+-+-+-+-+-+-+-+-+-+-+-+-+-+-+-+-+-+-+-+-+-+-+-+
|Version|  IHL  |Type of Service|          Total Length         |
+-+-+-+-+-+-+-+-+-+-+-+-+-+-+-+-+-+-+-+-+-+-+-+-+-+-+-+-+-+-+-+-+
|         Identification        |Flags|      Fragment Offset    |
+-+-+-+-+-+-+-+-+-+-+-+-+-+-+-+-+-+-+-+-+-+-+-+-+-+-+-+-+-+-+-+-+
|  Time to Live |    Protocol   |        Header Checksum         |
+-+-+-+-+-+-+-+-+-+-+-+-+-+-+-+-+-+-+-+-+-+-+-+-+-+-+-+-+-+-+-+-+
|                        Source Address                         |
+-+-+-+-+-+-+-+-+-+-+-+-+-+-+-+-+-+-+-+-+-+-+-+-+-+-+-+-+-+-+-+-+
|                      Destination Address                      |
+-+-+-+-+-+-+-+-+-+-+-+-+-+-+-+-+-+-+-+-+-+-+-+-+-+-+-+-+-+-+-+-+
|                    Options                 |     Padding       |
+-+-+-+-+-+-+-+-+-+-+-+-+-+-+-+-+-+-+-+-+-+-+-+-+-+-+-+-+-+-+-+-+
```

**IPv6 Header Structure**:

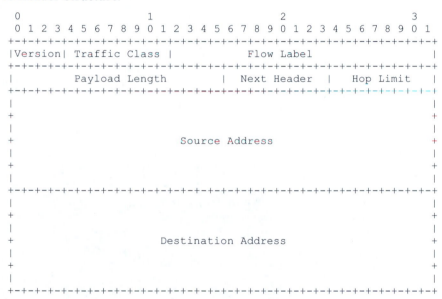

```
 0                   1                   2                   3
 0 1 2 3 4 5 6 7 8 9 0 1 2 3 4 5 6 7 8 9 0 1 2 3 4 5 6 7 8 9 0 1
+-+-+-+-+-+-+-+-+-+-+-+-+-+-+-+-+-+-+-+-+-+-+-+-+-+-+-+-+-+-+-+-+
|Version| Traffic Class |            Flow Label                 |
+-+-+-+-+-+-+-+-+-+-+-+-+-+-+-+-+-+-+-+-+-+-+-+-+-+-+-+-+-+-+-+-+
|         Payload Length        |  Next Header  |   Hop Limit    |
+-+-+-+-+-+-+-+-+-+-+-+-+-+-+-+-+-+-+-+-+-+-+-+-+-+-+-+-+-+-+-+-+
|                                                               |
+                                                               +
|                                                               |
+                      Source Address                           +
|                                                               |
+                                                               +
|                                                               |
+-+-+-+-+-+-+-+-+-+-+-+-+-+-+-+-+-+-+-+-+-+-+-+-+-+-+-+-+-+-+-+-+
|                                                               |
+                                                               +
|                                                               |
+                   Destination Address                         +
|                                                               |
+                                                               +
|                                                               |
+-+-+-+-+-+-+-+-+-+-+-+-+-+-+-+-+-+-+-+-+-+-+-+-+-+-+-+-+-+-+-+-+
```

**ICMP Header Structure:**

```
 0                   1                   2                   3
 0 1 2 3 4 5 6 7 8 9 0 1 2 3 4 5 6 7 8 9 0 1 2 3 4 5 6 7 8 9 0 1
+-+-+-+-+-+-+-+-+-+-+-+-+-+-+-+-+-+-+-+-+-+-+-+-+-+-+-+-+-+-+-+-+
|     Type      |     Code      |          Checksum             |
+-+-+-+-+-+-+-+-+-+-+-+-+-+-+-+-+-+-+-+-+-+-+-+-+-+-+-+-+-+-+-+-+
|                            unused                             |
+-+-+-+-+-+-+-+-+-+-+-+-+-+-+-+-+-+-+-+-+-+-+-+-+-+-+-+-+-+-+-+-+
|      Internet Header + 64 bits of Original Data Datagram      |
+-+-+-+-+-+-+-+-+-+-+-+-+-+-+-+-+-+-+-+-+-+-+-+-+-+-+-+-+-+-+-+-+
```

# APPENDIX K

**IPv4 to IPv6 Translation
with Fragmentation**

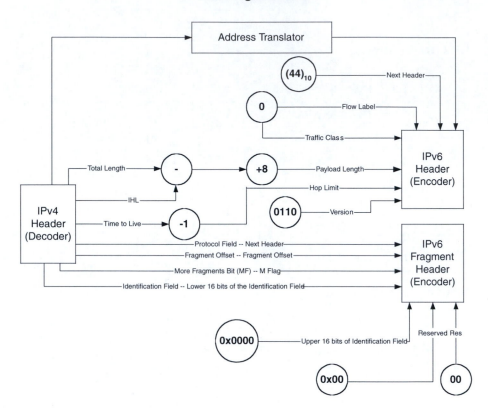

# APPENDIX L

**IPv4 to IPv6 Translation
without Fragmentation**

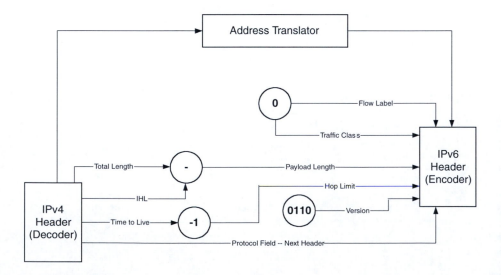

# APPENDIX M

### IPv6 to IPv4 Translation
### with Fragmentation

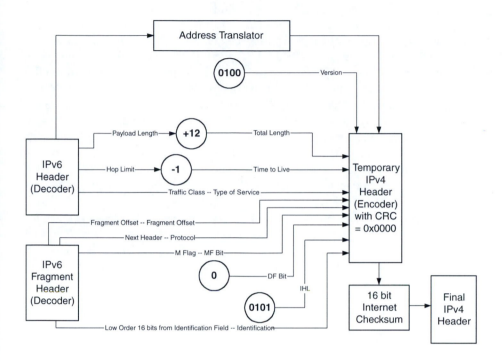

# APPENDIX N

### IPv6 to IPv4 Translation
### without Fragmentation

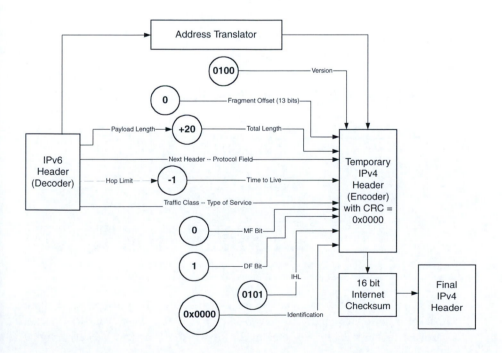

# APPENDIX O
# DESIGN SIMULATION

The senior design team generated individual test vectors that represent both fragmented and unfragmented IP traffic. This was done as an initial test to discover any inherent logical errors. Simulation output of these initial test vectors can be seen in the remainder of this appendix. To fully verify the design, a comprehensive simulation scheme was developed. The process had three parts: capturing IPv4 packets, converting the captured packet into IPv6 using our design and a software model, and comparing the results. A similar scheme was implemented to test the opposing pipeline. A software tool was developed that intercepted IPv4 packets from an active network. An explanation of this tool can be found in Appendix P. These intercepted packets were run through a software model of our device. It performs the same function as the bridge and generates the same IPv6 packet that our hardware bridge generates. The same packet was sent through our bridge and the two IPv6 packets were compared. If the two files were the same we could say that the bridge worked for that packet.

There is no IPv6 network that we could connect to in order to get data for the IPv6 to IPv4 pipeline. The IPv6 packets that were generated from the IPv4 to IPv6 pipeline were sent through the IPv6 to IPv4 pipeline as test data. IPv4 packets were outputted and were compared to the original IPv4 packets taken from the network. The two files were the same except for four fields: *Time to Live, Identification, Type of Service,* and *Checksum.* The *Time to Live* field had a value of 2 less than the original. It was found that since the field is decremented by one with each translation it goes through, the field should decrease by two after passing through the software model and then the pipeline. The fields of *Identification* and *Type of Service* were set to 0 in the IPv6 to IPv4 pipeline or in the IPv4 to IPv6 pipeline, making them different from the original packet. Since these three fields were different, the *Checksum,* which is calculated over the header, would be different between the two packets. This testing process was repeated with one hundred different IPv4 packets both fragmented and unfragmented. After running these simulation schemes, we found the pipelines functional and concluded our testing.

## SIMULATION DIAGRAMS, IPv4 TO IPv6

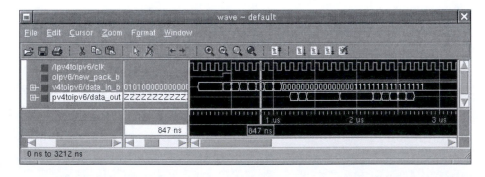

Simulation output
for an IPv4
fragmented packet.

The above simulation output was generated from a fragmented IPv4 input packet. The input signals are new_packet_b and data_in_b. The IPv4 packet arrives on the data_in_b signal in 32-bit words. The new_packet_b signal notifies our device that this is the start of a new packet. The data_out signal shows the output of our device as a fully translated IPv6 packet with 10 lines of header, 2 fragmentation lines, and the 3 lines of packet data that were passed to the device. Notice that the output begins 10 clock cycles after the first line of the input packet is received. This matches our theoretical pipeline timing diagrams in Appendices F–I. This scheme allows for packet translation without fear of data corruption.

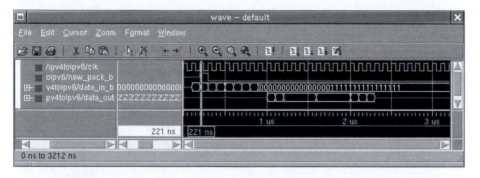

Simulation output
for an IPv4
unfragmented
packet.

   This simulation shows the translation of an IPv4 unfragmented packet. This simulation mirrors the input pattern of the above simulation so that will not be covered in detail. What is important to note is that the output packet, data_out, is comprised of 10 lines of header and 3 lines of data. Because this is an unfragmented packet, the stacked fragment header, 2 lines in IPv6, have not been generated.

## SIMULATION DIAGRAMS, IPv6 TO IPv4

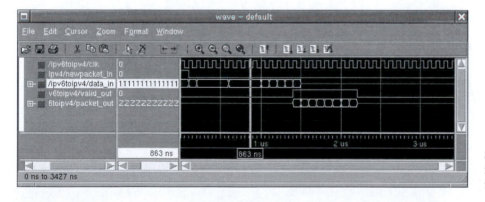

Simulation output
for an IPv6
fragmented packet.

The above simulation output was generated from a fragmented IPv6 input packet. The input signals are new_packet_b and data_in_b. The IPv6 packet arrives on the data_in_b signal in 32-bit words. The new_packet_b signal notifies our device that this is the start of a new packet. The data_out signal shows the output of our device as a fully translated IPv4 packet with 5 lines of header and the 3 lines of packet data that were passed to the device. Notice that the output begins on the third word of input packet data. This matches our theoretical pipeline timing diagrams in Appendices F–I. This scheme allows for packet translation without fear of data corruption. A valid_out signal has been added to the simulation to show when our device is providing valid output.

The simulation below shows the translation of an IPv6 unfragmented packet. This simulation mirrors the input pattern of the above simulation, except that the 2 line fragmentation header has been removed from our input pattern. The output is seen as a translated IPv4 unfragmented packet. The output looks similar to that of the previous simulation, but the two differ in the information in the header lines. IPv4 packets do not vary in structure, i.e., the number of lines, but change the information in the lines based on fragmentation.

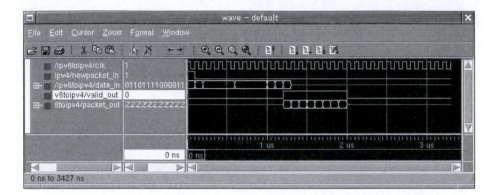

Simulation output
for an IPv6
unfragmented
packet.

## SIMULATION DIAGRAMS, IPv4 TO IPv6 WITH ICMP

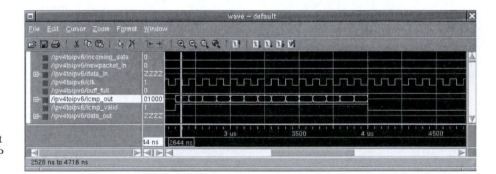

Simulation output
for an IPv4 ICMP
packet.

This simulation shows the generation of an IPv4 ICMP error packet due to an incorrect address in the packet header. This error condition was generated between two successfully translated packets to prove that the device could handle ICMP packet generation in the context of normal network traffic. As can be seen here the ICMP header consists of a standard IPv4 header followed by the extension header for the ICMP packet. These header formats can be seen in Appendix J. The payload of the packet contains the IP header and the first 64 bits of data of the offending packet. Since the ICMP functionality was integrated in the last term of the design cycle, only a few error packets were generated and verified using the pipeline with ICMP capabilities.

In order to test and interface the IPv4-IPv6 pipeline in real-world situations, a suite of programs was developed to mate the coreware pipeline with an IPv4 network. These tools are able to capture packets off the network, format the captured packet into data that is readable by the VHDL test bench, simulate the action of the IPv4-IPv6 pipeline, and re-inject the packets into the network. Additionally, a simple client-server package was created in order to generate the necessary traffic for the testing of the pipeline. This appendix provides a brief description of each program and the locations of the necessary 3rd-party libraries. The code is optimized for the Linux operating system, but should compile on any Unix-like operating system.

*capture* (capture.c) grabs packets from an Ethernet network and records the data into a format that is readable by the VHDL test bench that has been designed for the pipeline. This provides real-world data to be used in the simulation of the pipeline. This code requires the *libpcap* library, which can be found at http://www.tcpdump.org/release/libpcap-0.6.2.tar.gz.

*translate* (translate.c) is a software model of the IPv4-IPv6 pipeline. The code reads in the ASCII-formatted data required by the test bench, converts the captured IPv4 packets into IPv6 packets, then writes the new packets to a file for testing with the IPv6-IPv4 pipeline. This code requires no additional libraries than what is present on a standard Unix development environment.

*udp_send* (udp_send.c) and *udp_receive* (udp_receive.c) form a simple client-server pair for the generation and reception of UDP (User Datagram Protocol) packets. The client transmits a packet to the server containing the next four characters to be read in from the terminal (*stdin*). The server then receives this packet and displays the contents on the screen. Again, no additional libraries are required than what is present on a standard Unix development environment that supports socket programming constructs.

*udp_build* (udp_build.c) reads in the ASCII-formatted packets generated by the test bench and then injects the data into the IP-layer of the network. This program parses the output from the simulated pipeline, generates a custom IP header, and injects it back into the network. This program requires the *libnet* library, which is available at http://www.packetfactory.net/libnet/dist/libnet.tar.gz.

# APPENDIX Q
# PROPOSED TIMELINE

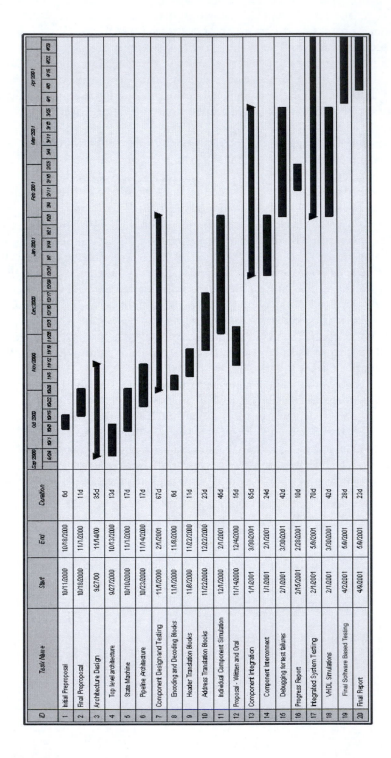

| ID | Task Name | Start | End | Duration |
|----|-----------|-------|-----|----------|
| 1 | Initial Preproposal | 10/11/2000 | 10/18/2000 | 6d |
| 2 | Final Preproposal | 10/18/2000 | 11/1/2000 | 11d |
| 3 | Architecture Design | 9/27/00 | 11/14/00 | 35d |
| 4 | Top level architecture | 9/27/2000 | 10/13/2000 | 13d |
| 5 | State Machine | 10/10/2000 | 11/1/2000 | 17d |
| 6 | Pipeline Architecture | 10/23/2000 | 11/14/2000 | 17d |
| 7 | Component Design and Testing | 11/1/2000 | 2/1/2001 | 67d |
| 8 | Encoding and Decoding Blocks | 11/1/2000 | 11/8/2000 | 6d |
| 9 | Header Translation Blocks | 11/8/2000 | 11/22/2000 | 11d |
| 10 | Address Translation Blocks | 11/22/2000 | 12/22/2000 | 23d |
| 11 | Individual Component Simulation | 12/1/2000 | 2/1/2001 | 46d |
| 12 | Proposal - Written and Oral | 11/14/2000 | 12/4/2000 | 15d |
| 13 | Component Integration | 1/1/2001 | 3/30/2001 | 65d |
| 14 | Component Interconnect | 1/1/2001 | 2/1/2001 | 24d |
| 15 | Debugging for test failures | 2/1/2001 | 3/30/2001 | 42d |
| 16 | Progress Report | 2/15/2001 | 2/28/2001 | 10d |
| 17 | Integrated System Testing | 2/1/2001 | 5/8/2001 | 70d |
| 18 | VHDL Simulations | 2/1/2001 | 3/30/2001 | 42d |
| 19 | Final Software Based Testing | 4/2/2001 | 5/9/2001 | 28d |
| 20 | Final Report | 4/9/2001 | 5/9/2001 | 23d |

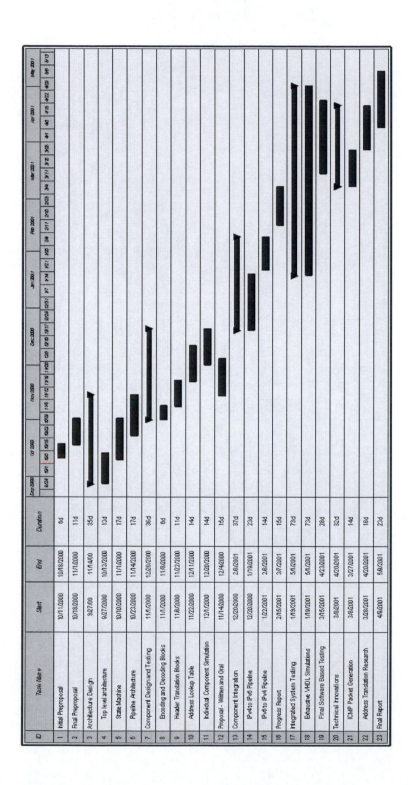

| ID | Task Name | Start | End | Duration |
|----|-----------|-------|-----|----------|
| 1 | Initial Preproposal | 10/11/2000 | 10/18/2000 | 6d |
| 2 | Final Preproposal | 10/18/2000 | 11/1/2000 | 11d |
| 3 | Architecture Design | 9/27/00 | 11/14/00 | 35d |
| 4 | Top level architecture | 9/27/2000 | 10/13/2000 | 13d |
| 5 | State Machine | 10/10/2000 | 11/1/2000 | 17d |
| 6 | Pipeline Architecture | 10/23/2000 | 11/14/2000 | 17d |
| 7 | Component Design and Testing | 11/1/2000 | 12/20/2000 | 36d |
| 8 | Encoding and Decoding Blocks | 11/1/2000 | 11/8/2000 | 6d |
| 9 | Header Translation Blocks | 11/8/2000 | 11/22/2000 | 11d |
| 10 | Address Lookup Table | 11/22/2000 | 12/11/2000 | 14d |
| 11 | Individual Component Simulation | 12/1/2000 | 12/20/2000 | 14d |
| 12 | Proposal - Written and Oral | 11/14/2000 | 12/4/2000 | 15d |
| 13 | Component Integration | 12/20/2000 | 2/6/2001 | 37d |
| 14 | IPv4 to IPv6 Pipeline | 12/20/2000 | 1/19/2001 | 23d |
| 15 | IPv6 to IPv4 Pipeline | 1/22/2001 | 2/8/2001 | 14d |
| 16 | Progress Report | 2/15/2001 | 3/7/2001 | 15d |
| 17 | Integrated System Testing | 1/19/2001 | 5/1/2001 | 73d |
| 18 | Exhaustive VHDL Simulations | 1/19/2001 | 5/1/2001 | 73d |
| 19 | Final Software Based Testing | 3/15/2001 | 4/23/2001 | 28d |
| 20 | Technical Innovations | 3/8/2001 | 4/20/2001 | 32d |
| 21 | ICMP Packet Generation | 3/8/2001 | 3/27/2001 | 14d |
| 22 | Address Translation Research | 3/28/2001 | 4/20/2001 | 18d |
| 23 | Final Report | 4/9/2001 | 5/9/2001 | 23d |

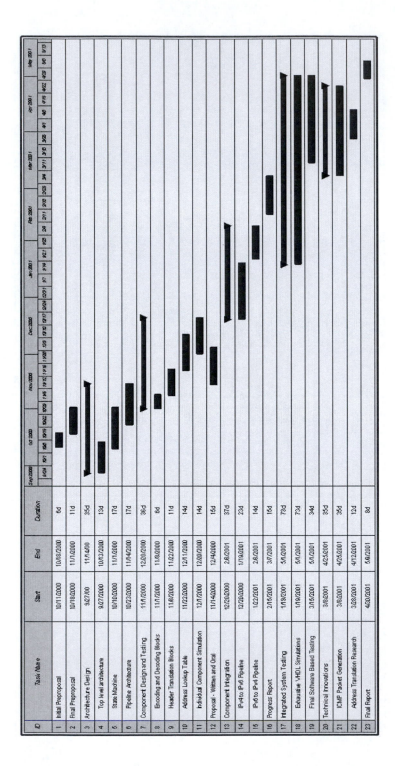

| ID | Task Name | Start | End | Duration |
|----|-----------|-------|-----|----------|
| 1 | Initial Preproposal | 10/11/2000 | 10/18/2000 | 6d |
| 2 | Final Preproposal | 10/18/2000 | 11/1/2000 | 11d |
| 3 | Architecture Design | 9/27/00 | 11/14/00 | 35d |
| 4 | Top level architecture | 9/27/2000 | 10/13/2000 | 13d |
| 5 | State Machine | 10/10/2000 | 11/1/2000 | 17d |
| 6 | Pipeline Architecture | 10/23/2000 | 11/14/2000 | 17d |
| 7 | Component Design and Testing | 11/1/2000 | 12/20/2000 | 36d |
| 8 | Encoding and Decoding Blocks | 11/1/2000 | 11/8/2000 | 6d |
| 9 | Header Translation Blocks | 11/8/2000 | 11/22/2000 | 11d |
| 10 | Address Lookup Table | 11/22/2000 | 12/11/2000 | 14d |
| 11 | Individual Component Simulation | 12/1/2000 | 12/20/2000 | 14d |
| 12 | Proposal - Written and Oral | 11/14/2000 | 12/4/2000 | 15d |
| 13 | Component Integration | 12/20/2000 | 2/6/2001 | 37d |
| 14 | IPv4 to IPv6 Pipeline | 12/20/2000 | 1/19/2001 | 23d |
| 15 | IPv6 to IPv4 Pipeline | 1/22/2001 | 2/6/2001 | 14d |
| 16 | Progress Report | 2/15/2001 | 3/7/2001 | 15d |
| 17 | Integrated System Testing | 1/19/2001 | 5/1/2001 | 73d |
| 18 | Exhaustive VHDL Simulations | 1/19/2001 | 5/1/2001 | 73d |
| 19 | Final Software Based Testing | 3/15/2001 | 5/1/2001 | 34d |
| 20 | Technical Innovations | 3/8/2001 | 4/25/2001 | 35d |
| 21 | ICMP Packet Generation | 3/8/2001 | 4/25/2001 | 35d |
| 22 | Address Translation Research | 3/28/2001 | 4/12/2001 | 12d |
| 23 | Final Report | 4/30/2001 | 5/9/2001 | 8d |

| Expense | Per Unit | # Needed | Cost |
|---|---|---|---|
| Software | | | |
| - ModelSim (per seat) | $10,000.00 | 4 | $40,000.00 |
| - Simplicity (per seat) | $20,000.00 | 4 | $80,000.00 |
| Desktop Systems | $2,000.00 | 4 | $8,000.00 |
| Networking Supplies | $100.00 | 1 | $100.00 |
| | | | |
| Total: | | | $128,100.00 |

Projected Expenditures for Design Team

| Expense | Per Unit | # Needed | Cost |
|---|---|---|---|
| Software | | | |
| - ModelSim (per seat) | $10,000.00 | 4 | $40,000.00 |
| - Simplicity (per seat) | $20,000.00 | 4 | $80,000.00 |
| Desktop Systems | $2,000.00 | 4 | $8,000.00 |
| Networking Supplies | $100.00 | 1 | $100.00 |
| Total | | | $128,100.00 |
| **Private Sector Expenses** | | | |
| Server | $4,000.00 | 1 | $4,000.00 |
| Engineer (9 month contract @ $80/hr) | $124,800.00 | 4 | $499,200.00 |
| Office Space (monthly) | $5,000.00 | 9 | $45,000.00 |
| Telecommunications (monthly) | $300.00 | 9 | $2,700.00 |
| | | | |
| Total: | | | $679,000.00 |

Projected Expenditures for Private-Sector Enterprise

# *Coreware* IPv4 to IPv6 Bridge
## Final Presentation

# Coreware IPv4 to IPv6 Bridge
## ECE-026

Keith Christman
Adam O'Donnell
Chayil Timmerman
Suma Varghese
Advisor: Dr. Harish Sethu

# Overview

- History and Future of the Internet Protocol
- Our Solution and Possible Alternatives
- Top Level Architecture
- Logical Subcomponents
- Pipeline Assembly
- Exhaustive Simulation
- Future Work
- Economic Analysis
- Social and Environmental Impact
- Demonstration

# The History and Future of the Internet Protocol

The Past: IPv4

- Exhausted Address Space
- Unnecessary Fields
- Security Concerns
- Configuration Overhead

The Future: IPv6

- Increased Address Space
- Efficient Header
- Enhanced Security
- Automated Configuration

# IPv4 Header

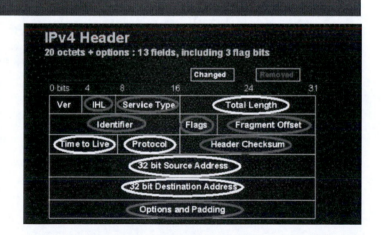

## IPv6 Header

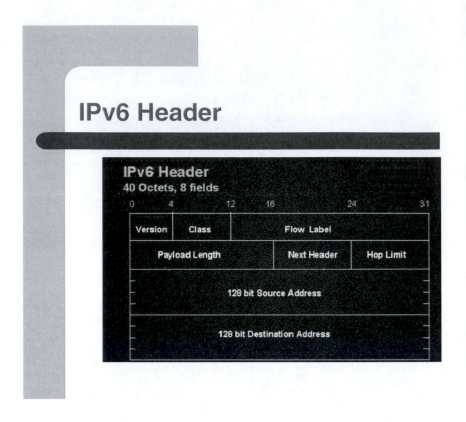

## Bumpy Road to the Future

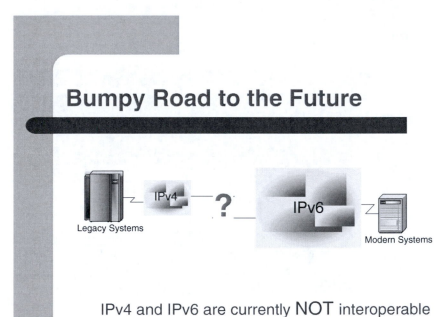

IPv4 and IPv6 are currently NOT interoperable

## Spanning the Divide

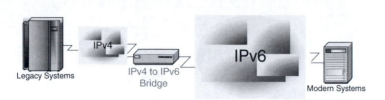

Bridging the Gap between the Protocols

## Our Solution

- Network Level Device
- Translator Style System
- Dual-Pipeline Architecture
- Coreware Description
  - Developed using VHDL

## Bridging Alternatives

- Host Level Software Solutions
- Network Level Software Solution
  - Open BSD
  - University of Washington
- Host Level Hardware Solutions

## Our Deliverable Solution

- Individual Logical Subcomponents
  - Decoder
  - Translation Blocks
  - Encoder
- Pipeline Construction
- ICMP Error Generation
- Comprehensive Simulation
- Software Support Tool
  - Demonstration

# Top Level Architecture

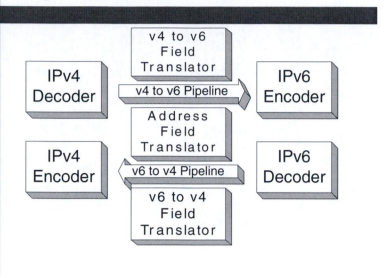

# Top Level Architecture

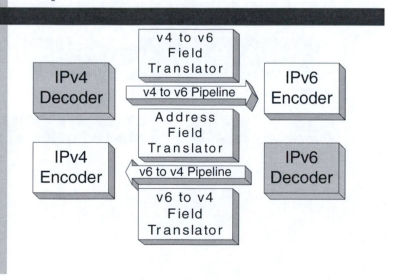

## Decoder Blocks

- 32-bit Information Bus
- Parses Incoming Packet Words
- Outputs Individual Protocol Fields
- Determines Start and End of a Packet

## IPv6 Decoder Block

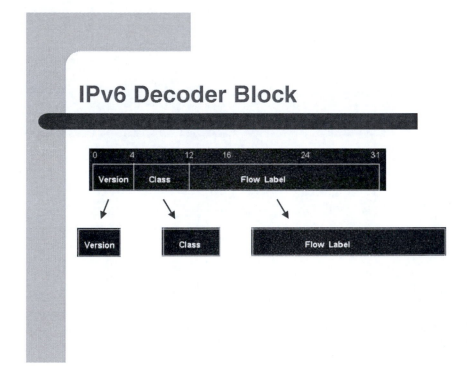

# Top Level Architecture

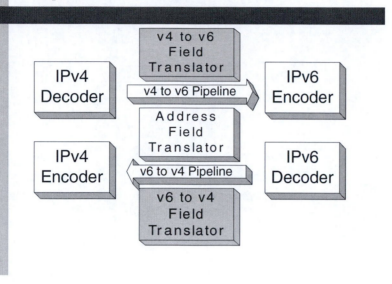

# Translation Logic

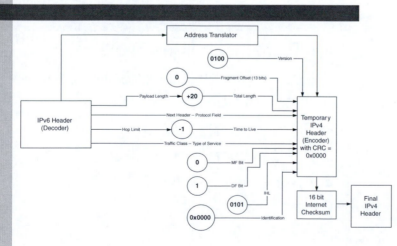

## Top Level Architecture

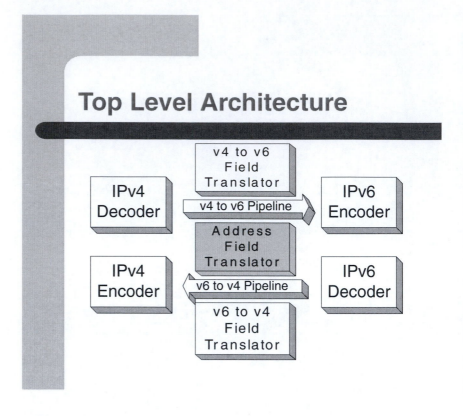

## Address Translation Logic

- 4 Element Look-up Table
- Router Manufacturer Supplies IPv4-IPv6 Address Pairs
- Placeholder Solution
  - Advancement of the design
- Area of Future Work

# Top Level Architecture

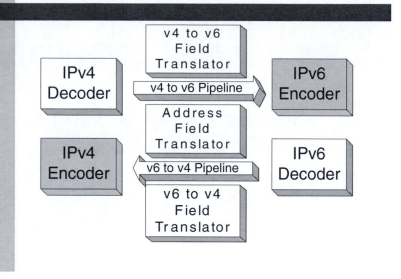

# Encoding Blocks

- Inverse operation of the decoder
- Compiles translated packet fields
- Outputs 32-bit words
- Signals the completed packet translation

# IPv6 Encoder Block

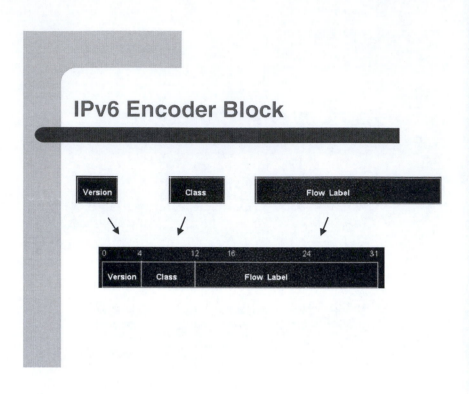

# Top Level Architecture

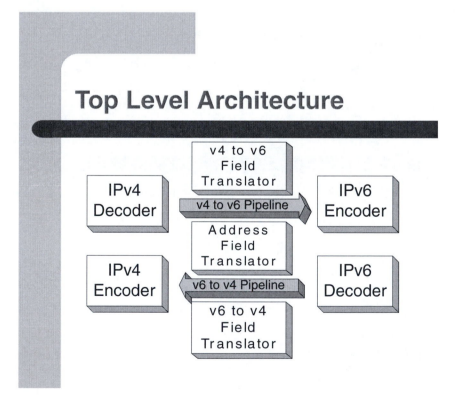

# Pipeline Assembly

- Interconnection of Logic Blocks
- Multiple Packets in Flight
- Timing of Sequential Packets
- Strategic Placement of Buffers
- Generation of Control Logic

# Pipeline Timing Diagram

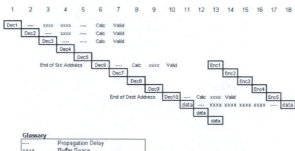

IPv6 to IPv4 Pipeline Timing Diagram for Unfragmented Packets

**Glossary**

| | |
|---|---|
| ---- | Propagation Delay |
| xxxx | Buffer Space |
| Dec | Decode Line Number |
| Enc | Encoder Line Number |
| Frag | Fragment Header |
| data/opt | Data or Options |
| Calc | Calculation translation block |

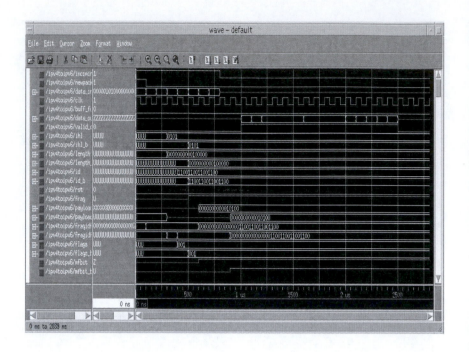

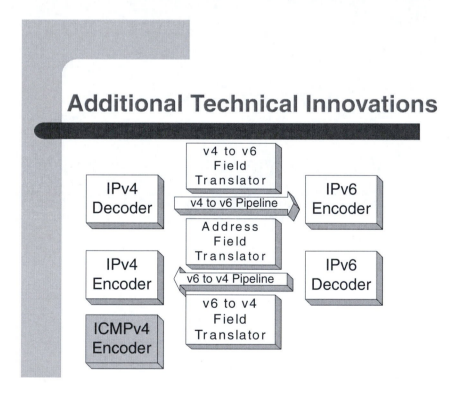

# ICMP Error Generation

- Internet Control Message Protocol
- Pipeline Detects Translation Errors
  - Source or Destination IP Addresses
  - Hop Limit or Time to Live Expires
- Save Offending Packet Data
- Added Ability to Stall the Pipeline
- Specialized ICMP Encoder

# Design Simulation

- Software tools
  - Capture IP Internet traffic
  - Software translation
- Fragmented and unfragmented IP traffic
- Variable length and sequence of packets
- Results verified for translated header and payload data

## Future Work

- Address Translation
  - Third Pipeline (DNS Capturing)
  - Add a Dictionary Chip or Searchable Memory Core
  - Memory I/O Currently Implemented
  - Adaptable Pipeline Architecture
- Testing with Randomized IP Traffic
- ICMP Error Detection for IPv6
- ICMP Exhaustive Simulation

## Gantt Chart

| ID | Task Name | Start | End | Duration | Q4 00 | | | Q1 01 | | | Q2 01 | |
|----|-----------|-------|-----|----------|-------|-----|-----|-----|-----|-----|-----|-----|
| | | | | | Oct | Nov | Dec | Jan | Feb | Mar | Apr | May |
| 1 | Initial Preproposal | 10/11/2000 | 10/18/2000 | 1.20w | | | | | | | | |
| 2 | Final Preproposal | 10/18/2000 | 11/1/2000 | 2.20w | | | | | | | | |
| 3 | Architecture Design | 9/27/2000 | 11/14/2000 | 7w | | | | | | | | |
| 4 | Component Design and Testing | 11/1/2000 | 12/20/2000 | 7.20w | | | | | | | | |
| 5 | Proposal - Written and Oral | 11/14/2000 | 12/4/2000 | 3w | | | | | | | | |
| 6 | Component Integration | 12/20/2000 | 2/8/2001 | 7.40w | | | | | | | | |
| 7 | Progress Report | 2/15/2001 | 3/7/2001 | 3w | | | | | | | | |
| 8 | Integrated System Testing | 1/19/2001 | 5/1/2001 | 14.60w | | | | | | | | |
| 9 | Technical Innovations | 3/8/2001 | 4/20/2001 | 6.40w | | | | | | | | |
| 10 | Final Report | 4/9/2001 | 5/9/2001 | 4.60w | | | | | | | | |

# Projected Expenditures

| Expense | Per Unit | # Needed | Cost |
|---|---|---|---|
| Software | | | |
| - ModelSim (per seat) | $10,000.00 | 4 | $40,000.00 |
| - Simplicity (per seat) | $20,000.00 | 4 | $80,000.00 |
| Desktop Systems | $2,000.00 | 4 | $8,000.00 |
| Networking Supplies | $100.00 | 1 | $100.00 |
| Total | | | $128,100.00 |
| **Private Sector Expenses** | | | |
| Server | $4,000.00 | 1 | $4,000.00 |
| Engineer (9 month contract @ $80/hr) | $124,800.00 | 4 | $499,200.00 |
| Office Space (monthly) | $5,000.00 | 9 | $45,000.00 |
| Telecommunications (monthly) | $300.00 | 9 | $2,700.00 |
| | | | |
| Total: | | | $679,000.00 |

# Economic Analysis

- Cost
  - Tuition
  - Opportunity cost of team's time
  - No manufacturing expenditures
- Sales and Profit
  - Profits based on end product
  - Licensing of intellectual property

## Social and Environmental Impacts

- Increased Availability of the Internet
- Extended Lifetime of Older Systems
- Reduction of Paper Consumption

## Now For a Demonstration

Software Support Tools:

Capturing and Translating **Live** Internet Traffic

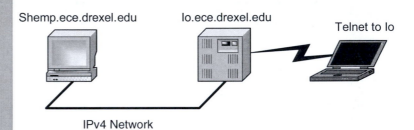

Shemp.ece.drexel.edu     Io.ece.drexel.edu     Telnet to Io

IPv4 Network

# Questions and Comments

Keith Christman     st96jq43@drexel.edu

Adam O'Donnell     adam@io.ece.drexel.edu

Chayil Timmerman     chayil@ieee.org

Suma Varghese     suma@ieee.org

# IPv4 to IPv6 with Fragmentation

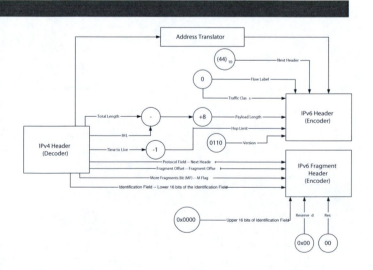

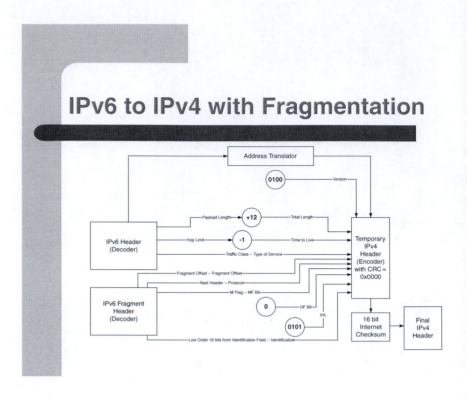

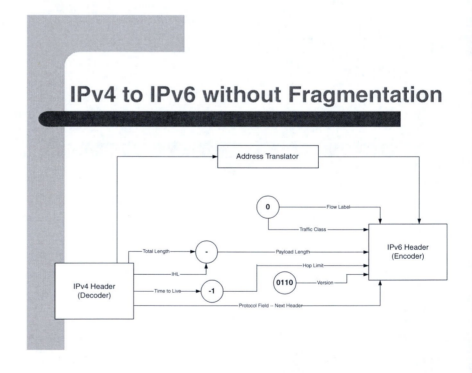

# References

IPv6: The Next Generation Internet! (10/18/2000). IPv6: Networking for the 21 st Century.
http://www.ipv6.org. [11/07/2000].

IP Next Generation Overview. (05/14/1995). IP Next Generation (IPng)
http://playground.sun.com/pub/ipng/html/INET-IPng-Paper.htm l. [11/07/2000]

RFC 791 – Internet Protocol. (09/1981). Internet RFC/STD/FYI/BCP Archives.
http://www.faqs.org/rfcs/rfc791.html. [09/26/2000].

RFC 2460 – Internet Protocol, Version 6 Specification (12/1998). Internet RFC/STD/FYI/BCP
Archives. http://www.faqs.org/rfcs/rfc2460.html. [09/26/2000].

RFC 792 – Internet Control Message Protocol (09/1981). Internet RFC/STD/FYI/BCP Archives.
http://www.faqs.org/rfcs/rfc792.html. [09/26/2000] .

RFC 2463 - Internet Control Message Protocol (ICMPv6) for the Internet Protocol Version 6 (IPv6)
Specification. (12/1998). Internet RFC/STD/FYI/BCP Archives.
http://www.faqs.org/rfcs/rfc2463.html. [09/26/2000].

J. Bhasker, *A VHDL Primer,* Upper Saddle River: Prentice-Hall, 1999.

RFC 2765 – Stateless IP/ICMP Translation Algorithm (SIIT). (02/2000). Internet RFC/STD/FYI/BCP
Archives. http://www.faqs.org/rfcs/rfc2765.html . [10/03/2000].

The Design and Implementation of an IPv6/IPv4 Network Address and Protocol Translator.
(06/1998). Brian N. Bershad [Personal Homepage].
http://www.cs.washington.edu/homes/bershad/Papers/USENIX98/nap.html . [10/03/2000].

# INDEX